TCP/IP Essentials

The TCP/IP family of protocols have become the de facto standard in the world of networking, are found in virtually all computer communication systems, and form the basis of today's Internet. *TCP/IP Essentials* is a hands-on guide to TCP/IP technologies, and shows how the protocols operate in practice. The book contains a series of carefully designed and extensively tested laboratory experiments that span the various elements of protocol definition and behavior. Topics covered include bridges, routers, LANs, static and dynamic routing, multicast and realtime service, and network management and security. The experiments are described in a Linux environment, with parallel notes on Solaris implementation. The book includes many exercises, and supplementary material for instructors is available. The book is aimed at students of electrical and computer engineering or computer science who are taking courses in networking. It is also an ideal guide for engineers studying for networking certifications.

SHIVENDRA S. PANWAR is a professor in the Electrical and Computer Engineering Department at Polytechnic University, Brooklyn, New York, USA. He is currently the Director of the New York State Center for Advanced Technology in Telecommunications (CATT). He is the author of over 80 refereed papers.

SHIWEN MAO is a research associate in the Bradley Department of Electrical and Computer Engineering, Virginia Polytechnic Institute and State University, Blacksburg, VA, USA.

JEONG-DONG RYOO is a senior member of research staff at the Electronics and Telecommunications Research Institute, Daejon, South Korea.

YIHAN LI is a research associate in the Department of Electrical Engineering, Polytechnic University, Brooklyn, New York, USA.

TCP/IP Essentials

A Lab-Based Approach

Shivendra S. Panwar

Department of Electrical and Computer Engineering,
Polytechnic University, Brooklyn, New York

Shiwen Mao

The Bradley Department of Electrical and Computer Engineering,
Virginia Polytechnic Institute and State University
Blacksburg, Virginia

Jeong-dong Ryoo

Electronics and Telecommunications Research Unit,
Daejeon, South Korea

Yihan Li

Department of Electrical and Computer Engineering,
Polytechnic University,
Brooklyn, New York

CAMBRIDGE
UNIVERSITY PRESS

PUBLISHED BY THE PRESS SYNDICATE OF THE UNIVERSITY OF CAMBRIDGE
The Pitt Building, Trumpington Street, Cambridge, United Kingdom

CAMBRIDGE UNIVERSITY PRESS
The Edinburgh Building, Cambridge CB2 2RU, UK
40 West 20th Street, New York, NY 10011-4211, USA
477 Williamstown Road, Port Melbourne, VIC 3207, Australia
Ruiz de Alarcón 13, 28014 Madrid, Spain
Dock House, The Waterfront, Cape Town 8001, South Africa

http://www.cambridge.org

First published 2004

Printed in the United Kingdom at the University Press, Cambridge

Typefaces Times 11/14.5 pt. and Helvetica   *System* LATEX 2$_\varepsilon$ [TB]

A catalog record for this book is available from the British Library

Library of Congress Cataloging in Publication data

TCP/IP essentials : a lab-based approach / Shivendra S. Panwar . . . [et al.]
     p.   cm.
   Includes bibliographical references and index.
   ISBN 0 521 84144 5 (hardback)   ISBN 0 521 60124 X (paperback)
   1. TCP/IP (Computer network protocol)   2. Computer networks.   I. Panwar, Shivendra S.

   TK5105.585.T355 2004
   004.6′2–dc22   2004045664

ISBN 0 521 84144 5 hardback
ISBN 0 521 60124 X paperback

To my wife, Shruti, my parents, and Choti.

Shivendra Panwar

To my wife, Kweesook, my children, James and Michelle, and my parents.

Jeong-dong Ryoo

To our son, Eric, and our parents.

Yihan Li and Shiwen Mao

Contents

Preface

You can know the name of a bird in all the languages of the world, but when you're finished, you'll know absolutely nothing whatever about the bird . . . So let's look at the bird and see what it's doing – that's what counts. I learned very early the difference between knowing the name of something and knowing something.
Richard Feynman (1918–1988)

As the title of this book suggests, this book is a *minimalist* approach to teaching TCP/IP *using laboratory-based experiments*. It is minimalist in that it provides one, possibly idiosyncratic, choice of topics at a depth we felt was sufficient to learn the basics of TCP/IP. The intention was not to write a reference text on the subject. The laboratory was important in giving students the experience of observing the TCP/IP protocols in action. The act of observing and drawing some conclusions from those observations, brings to life the often dry study of network protocols, and motivates students to learn more about them.

Appendix A is necessary reading only for the instructor who is in charge of setting up the lab. We have attempted to keep costs down so that only the most Scrooge-like University administrator would raise an eyebrow over the cost of the lab equipment (as for lab space, that may be another matter!). We assume that the students have a basic background in networking, perhaps from a previous course, or perhaps as part of a course that back loads the experiments in this book after providing a general lecture-based introduction to networks. Chapter 0 is a quick overview of TCP/IP that serves two purposes. It provides an overview of the TCP/IP stack, and serves as the framework for the rest of the book. Chapters 1 to 9 have the following common structure. Each of them provides introductory material suitable for presentation in the lecture part of the course followed by a lab experiment. The lab experiments should follow lectures that provide the students with the basic knowledge they need to perform the experiments and derive insights from their observations during the course of the experiments. Each lab experiment is designed to take no more than 3 hours to complete.

The experiments were developed on the basis of a course taught at the Polytechnic University over the course of over eight years. Initially, we used SUN workstations with the Solaris operating system, but have now switched to Linux machines. The primary operating system in this book is Linux, but with Solaris commands provided when they differ from Linux commands. Chapter 1 provides an introduction to Linux, since many students may be unfamiliar with this operating system. It also introduces key tools used in subsequent experiments such as tcpdump and Ethereal. Chapter 2 introduces network interfaces, ping and IP addresses. Chapter 3 introduces bridges, also known as layer two switches, bridge/router configuration, and the Cisco IOS. Chapter 4 focuses on routing, with RIP and OSPF as the routing protocols studied, along with the useful traceroute utility. Chapter 5 introduces UDP and FTP. Chapter 6 follows up with TCP, including a study of its congestion control mechanism. These six chapters are sufficient in many cases to introduce students to the basics of TCP/IP. Nonetheless, the next three chapters are important for students who wish to link the basic plumbing of TCP/IP with applications. Chapter 7 deals with IP multicast and realtime applications. The web, DHCP, NTP and NAT are some key applications that are presented in Chapter 8, as well as a brief introduction to socket programming. Network management and security are arguably two of the most important features that students need to know, at least at a basic level. Chapter 9 provides a brief introduction to this material, which can easily be the subject of a separate course. A list of key RFCs is provided at the end of the book.

There are several alternative ways of teaching this material with this book. A general knowledge of networking is assumed as a prerequisite for this book. However, an introductory course in networking could be combined with the first six experiments, back-loaded at the end of the course, to illustrate the lowest four layers of the protocol stack. For computer scientists, a top-down approach is sometimes the preferred approach in teaching networking. In that case the lab experiments can be re-ordered to focus on the higher layers.

Note to instructors

Additional course material, including lecture transparencies, sample lab reports, homework assignments, examinations, and errata, are available at the course website: www.cambridge.org/052160124X.

Acknowledgements

The authors would like to acknowledge the support of Polytechnic University, the National Science Foundation, the New York State Office of Science, Technology and Academic Research (NYSTAR), and the Securities Industries Automation Corporation (SIAC). In particular, it was our work with SIAC, a company responsible for the networking and system needs of the New York and American Stock Exchanges, which initially inspired us. In particular, we would like to thank Andrew Bach, Joseph Kubat, Michael Lamberg, Darko Mrakovcic, and Dror Segal of SIAC for their support. A special thanks to Dr. Nitin Gogate, who helped with the initial version of the experiments, and all the graduate students who followed. We would like to thank Jeffrey (Zhifeng) Tao, Yanming Shen and Pei Liu, who helped proofread and test the lab experiments. We would also like to thank the following faculty members who have also taught this course over the years at Poly: Malathi Veeraraghavan, John (Zheng-Xue) Zhao, and Jorg Liebeherr.

General conventions

The following conventions are used all through this book.

- In paragraphs, Linux, Unix and Cisco IOS commands are written in a bold font, such as: **telnet** and **enable**.
- In a compound command with options and parameters, the command and options are in bold, while the parameters are in italics. For example, in

  tcpdump -enx host *ip_addr1* **and** *ip_addr2*,

  the command **tcpdump** uses options **-e, -n** and **-x**. In the filter that follows, key words such as **host, and, not, or** etc., are also in bold. The parameters are *ip_addr1* and *ip_addr2*, which should be replaced with the corresponding IP addresses during the exercise.
  The following exemplary command,

  /etc/init.d/snmpd start|stop,

  uses two options. Either **start** or **stop** can be used, but not at the same time.
- The name of a host or router is in the `Typewriter` typestyle, e.g., `shakti` or `Router4`.
- A protocol header field is also in the `Typewriter` typestyle, e.g., `Length` or `Source IP Address`.
- Questions in the LAB REPORT section of each exercise should be answered in the lab report. For example, for Exercise 1 in Chapter 1, students need to answer the following question in Lab Report 1.
  LAB REPORT What is the default directory when you open a new command tool? What is your working directory?
- In this guide, we focus on the Linux operating system. However, this guide can also be used with the Sun Solaris operating system. In the following text, Linux-specific material, or general material that apply to both operating systems are used, while the Solaris specific materials are enclosed between horizontal lines.

Abbreviations

ACK	Acknowledgement
AIMD	Additive-Increase-Multiplicative-Decrease
API	Application Programming Interface
ARP	Address Resolution Protocol
ARPA	Advanced Research Projects Agency
API	Application Programming Interface
AS	Autonomous System
ATM	Asynchronous Transfer Mode
BGP	Border Gateway Protocol
BOOTP	Bootstrap Protocol
BPDU	Bridge Protocol Data Unit
BSD	Berkely Software Distribution
CDE	Common Desktop Environment
CIDR	Classless Interdomain Routing
CBT	Core-Based Tree
CGI	Common Gateway Interface
CRC	Cyclic Redundancy Check
CSMA/CA	Carrier Sense Multiple Access/Collision Avoidance
CSMA/CD	Carrier Sense Multiple Access/Collision Detection
DBS	Distributed Benchmark System
DES	Data Encryption Standard
DHCP	Dynamic Host Configuration Protocol
DNS	Domain Name System
DSS	Digital Signature Standard
DVMRP	Distance Vector Multicast Routing Protocol

EGP	Exterior Gateway Protocol
FDDI	Fiber Distributed Data Interface
FEC	Forward Error Correction
FIN	Finish Flag
FTP	File Transfer Protocol
GPS	Global Positioning System
HTML	HyperText Markup Language
HTTP	HyperText Transfer Protocol
IAB	Internet Architecture Board
ICANN	Internet Corporation for Assigned Names and Numbers
ICMP	Internet Control Message Protocol
IETF	Internet Engineering Task Force
IGP	Interior Gateway Protocol
IGMP	Internet Group Management Protocol
InterNIC	Internet Network Information Center
IP	Internet Protocol
IRTF	Internet Research Task Force
ISOC	Internet Society
ISN	Initial Sequence Number
LAN	Local Area Network
LSA	Link State Advertisement
MAC	Medium Access Control
MAC	Message Authentication Code
MIB	Management Information Base
MOSPF	Multicast Extension to OSPF
MPLS	Multiprotocol Label Switching
MSL	Maximum Segment Life
MSS	Maximum Segment Size
MTU	Maximum Transmission Unit

NAT	Network Address Translator
NFS	Network File System
NIST	National Institute of Standards and Technology
NTP	Network Time Protocol
OSPF	Open Shortest Path First
PAT	Port Address Translation
PDA	Personal Digital Assistant
PDU	Protocol Data Unit
PIM	Protocol Independent Multicast
PNG	Portable Network Graphics
PPP	Point-to-Point Protocol
QoS	Quality of Service
RIP	Routing Information Protocol
RARP	Reverse Address Resolution Protocol
RBAC	Role-Based Access Control
RFC	Request for Comments
RPC	Remote Procedure Call
RRQ	Read Request
RSA	Rivest–Shamir–Adleman
RST	Reset Flag
RTO	Retransmission Timeout
RTCP	Realtime Transport Control Protocol
RTP	Realtime Transport Protocol
RTSP	Real Time Streaming Protocol
RTT	Round-Trip Time
SACK	Selective Acknowledgment
SHA	Secure Hash Algorithm
SIP	Session Initiation Protocol
SMI	Structure of Management Information
SMTP	Simple Mail Transfer Protocol
SNMP	Simple Network Management Protocol
SPF	Shortest Path First

SSL	Secure Sockets Layer
STDIN	Standard Input
STDOUT	Standard Output
SYN	Synchronize Sequence Number Flag
TCP	Transmission Control Protocol
TE	Traffic Engineering
TFTP	Trivial File Transfer Protocol
TTL	Time-to-Live
UDP	User Datagram Protocol
UI	User Interface
VoIP	Voice over IP
VPN	Virtual Private Network
WAN	Wide Area Network
Wi-Fi	Wireless Fidelity
WWW	World Wide Web

0 TCP/IP overview

From these assumptions comes the fundamental structure of the Internet: a packet switched communications facility in which a number of distinguishable networks are connected together using packet communications processors called gateways which implement a store and forward packet forwarding algorithm.

David D. Clark

0.1 The Internet

The Internet is a global information system consisting of millions of computer networks around the world. Users of the Internet can exchange email, access to the resources on a remote computer, browse web pages, stream live video or audio, and publish information for other users. With the evolution of *e-commerce*, many companies are providing services over the Internet, such as on-line banking, financial transactions, shopping, and on-line auctions. In parallel with the expansion in services provided, there has been an exponential increase in the size of the Internet. In addition, various types of electronic devices are being connected to the Internet, such as cell phones, personal digital assistants (PDA), and even TVs and refrigerators.

Today's Internet evolved from the ARPANET sponsored by the Advanced Research Projects Agency (ARPA) in the late 1960s with only four nodes. The Transmission Control Protocol/Internet Protocol (TCP/IP) protocol suite, first proposed by Cerf and Kahn in [1], was adopted for the ARPANET in 1983. In 1984, NSF funded a TCP/IP based backbone network, called NSFNET, which became the successor of the ARPANET. The Internet became completely commercial in 1995. The term "Internet" is now used to refer to the global computer network loosely connected together using packet switching technology and based on the TCP/IP protocol suite.

The Internet is administered by a number of groups. These groups control the TCP/IP protocols, develop and approve new standards, and assign Internet addresses and other resources. Some of the groups are listed here.

- Internet Society (ISOC). This is a professional membership organization of Internet experts that comments on policies and practices, and oversees a number of other boards and task forces dealing with network policy issues.
- Internet Architecture Board (IAB). The IAB is responsible for defining the overall architecture of the Internet, providing guidance and broad direction to the IETF (see below).
- Internet Engineering Task Force (IETF). The IETF is responsible for protocol engineering and development.
- Internet Research Task Force (IRTF). The IRTF is responsible for focused, long-term research.
- Internet Corporation for Assigned Names and Numbers (ICANN). The ICANN has responsibility for Internet Protocol (IP) address space allocation, protocol identifier assignment, generic and country code Top-Level Domain name system management, and root server system management functions. These services were originally performed by the Internet Assigned Numbers Authority (IANA) and other entities. ICANN now performs the IANA function.
- Internet Network Information Center (InterNIC). The InterNIC is operated by ICANN to provide information regarding Internet domain name registration services.

The Internet standards are published as *Request for Comments* (RFC), in order to emphasize the point that "the basic ground rules were that anyone could say anything and that nothing was official" [2]. All RFCs are available at the IETF's website http://www.ietf.org/. Usually, a new technology is first proposed as an *Internet Draft*, which expires in six months. If the Internet Draft gains continuous interest and support from ISOC or the industry, it will be promoted to a RFC, then to a *Proposed Standard*, and then a *Draft Standard*. Finally, if the proposal passes all the tests, it will be published as an *Internet Standard* by IAB.

0.2 TCP/IP protocols

The task of information exchange between computers consists of various functions and has tremendous complexity. It is impractical, if not

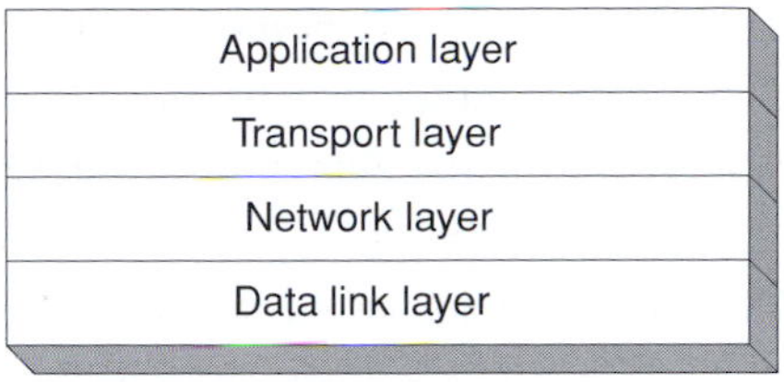

| Application layer |
| Transport layer |
| Network layer |
| Data link layer |

Figure 0.1. The TCP/IP protocol stack.

impossible, to implement all these functions in a single module. Instead, a *divide-and-conquer* approach was adopted. The communication task is broken up into subtasks and organized in a hierarchical way according to their dependencies to each other. More specifically, the subtasks, each of which is responsible for a facet of communication, are organized into different layers. Each higher layer uses the service provided by its lower layers, and provides service to the layers above it. The service is provided to the higher layer transparently, while heterogeneity and details are hidden from the higher layers. A protocol is used for communication between entities in different systems, which typically defines the operation of a subtask within a layer.

TCP/IP protocols, also known more formally as the *Internet Protocol Suite*, facilitates communications across interconnected, heterogeneous computer networks. It is a combination of different protocols, which are normally organized into four layers as shown in Fig. 0.1. The responsibility and relevant protocols at each layer are now given.

• The *application layer* consists of a wide variety of applications, among which are the following.
  • Hypertext Transfer Protocol (HTTP). Provides the World Wide Web (WWW) service.
  • Telnet. Used for remote access to a computer.
  • Domain Name System (DNS). Distributed service that translates between domain names and IP addresses.
  • Simple Network Management Protocol (SNMP). A protocol used for managing network devices, locally or remotely.
  • Dynamic Host Configuration Protocol (DHCP). A protocol automating the configuration of network interfaces.
• The *transport layer* provides data transport for the application layer, including the following.
  • Transmission Control Protocol (TCP). Provides *reliable* data transmission by means of *connection-oriented* data delivery over an IP network.

- User Datagram Protocol (UDP). A *connectionless* protocol, which is simpler than TCP and does not guarantee reliability.
- The *network layer* handles routing of packets across the networks, including the following.
  - Internet Protocol (IP). The "workhorse" of the TCP/IP protocol stack, which provides *unreliable* and *connectionless* service.
  - Internet Control Message Protocol (ICMP). Used for error and control messages.
  - Internet Group Management Protocol (IGMP). Used for *multicast* membership management.
- The *link layer* handles all the hardware details to provide data transmission for the network layer. Network layer protocols can be supported by various link layer technologies, such as those listed here.
  - Ethernet. A popular *multiple access* local area network protocol.
  - Wireless LAN. A wireless multiple access local area network based the IEEE 802.11 standards.
  - Point to Point Protocol (PPP). A *point-to-point* protocol connecting pairs of hosts.
  - Address Resolution Protocol (ARP). Responsible for resolving network layer addresses.

Figure 0.2 shows the relationship among protocols in different layers. We will discuss these protocols in more detail in later chapters.

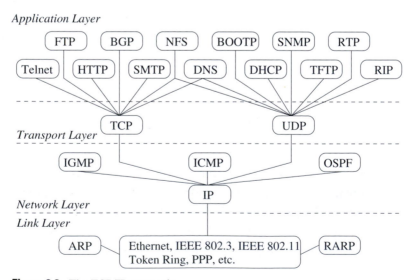

Figure 0.2. The TCP/IP protocols.

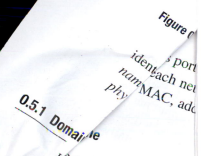

Figure 0.3. An illustration of the layers involved when two hosts communicate over the same Ethernet segment or over an Ethernet hub.

0.3 Internetworking devices

The Internet is a collection of computers connected by internetworking devices. According to their functionality and the layers at which they are operating, such devices can be classified as *hubs*, *bridges*, *switches*, and *routers*.

Hubs are physical layer devices, used to connect multiple hosts. A hub simply copies frames received from a port to all other ports, thus emulating a broadcast medium. Bridges, sometimes called *layer two switches*,[1] are link layer devices. They do not examine upper layer information, and can therefore forward traffic rapidly. Bridges can be used to connect distant stations and thus extend the effective size of a network. Bridges are further discussed in Chapter 3.

Routers, also called *layer three switches*, are network layer devices incorporating the routing function. Each router maintains a *routing table*, each entry of which contains a destination address and a next-hop address. None of the routers has information for the complete route to a destination. When a packet arrives, the router checks its routing table for an entry that matches the destination address, and then forwards the packet to the next-hop address. Routing is further discussed in Chapter 4.

Figure 0.3 shows the layers involved in communication between two hosts when they are connected by an Ethernet hub. The hosts can directly

---

[1] The industry, confusingly, also uses the term *smart hubs* for switches.

Table 0.1. Classification of the generic do

domain	Description
com	Commercial organization
edu	Educational institutions
gov	other US government institu
int	International organizations
mil	U.S. military groups
net	Major network support centers
org	Other organizations

top-level domains, .aero, .biz, .coop, .info, .museu d .pro, were
added to the Internet's domain name system by IC 0.
Since the TCP/IP programs only recognize num main name
system (DNS) is used to resolve, i.e., translate, a ame to the
corresponding IP address. Then the resolved IP add er than the
domain name, is used in the TCP/IP kernel. DNS is server type
of service. Since the entire database of domain names a ddresses is
too large for any single server, it is implemented as dist databases
maintained by a large number of DNS servers (usually hos puters running the DNS server program). Thus each DNS server or aintains a
portion of the domain name database shown in Fig. 0.8. A l can query
the DNS servers for the IP address associated with a domai e, or for
the domain name associated with an IP address. If the DNS ser being
queried does not have the target entry in its database, it may cont t other
DNS servers for assistance. Or, it may returns a list of other DNS ervers
that may contain the information. Thus the client can query these servers
iteratively.

It is inefficient to perform name resolution for the same domain name
every time its IP address is requested. Instead, DNS servers and clients
use *name caching* to reduce the number of such queries. A DNS server
or client maintains a cache for the names and corresponding IP addresses
which have been recently resolved. If the requested domain name is in the
cache, then there is no need to send a DNS query to resolve it. In addition,
each cached entry is associated with a Time-to-Live timer. The value of
this timer, which is usually set to the number of seconds in two days when
the entry is first cached, is determined by the server that returns the DNS
reply. The entry will be removed from the cache when the timer expires.

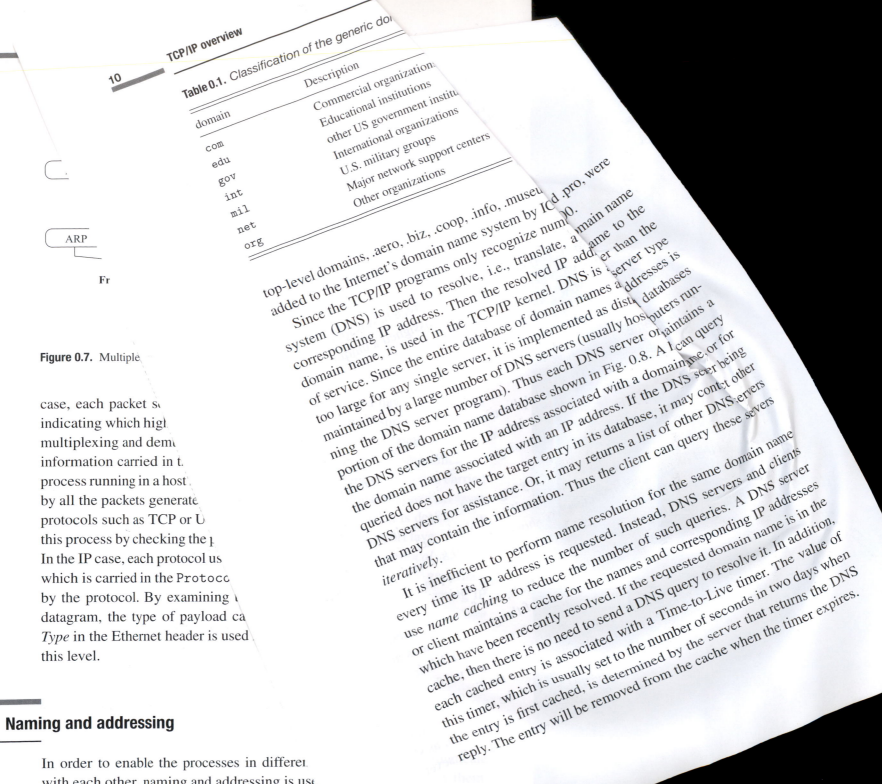

ARP

Fr

Figure 0.7. Multiple

case, each packet s
indicating which high
multiplexing and dem
information carried in t.
process running in a host
by all the packets generate
protocols such as TCP or U
this process by checking the
In the IP case, each protocol us
which is carried in the Protocc
by the protocol. By examining
datagram, the type of payload ca
Type in the Ethernet header is used
this level.

0.5 Naming and addressing

In order to enable the processes in differen
with each other, naming and addressing is use
As discussed in the previous section, a proces

0.5.2 Port number

Port numbers are used as addresses for application layer user processes. The value of the `Port Number` field in the TCP or UDP header is used to decide which application process the data belongs to.

Most network applications are implemented in a client–server architecture, where a server provides a service to the network users, and a client requests the service from the server. The server is always running and uses a *well-known* port number. Well-known port numbers from 1 to 255 are used for Internet-wide services (e.g., **telnet** uses 23 and **ssh** uses port 22), while those from 256 to 1023 are preserved for Unix specific services (e.g., **rlogin** uses 513). On the other hand, a client runs for a period of time associated with the time needed to fullfil its request. It starts up, sends requests to the server, receives service from the server, and then terminates. Therefore clients use *ephemeral* port numbers which are randomly chosen and are larger than 1023.

0.5.3 IP address

Each host interface in the Internet has a unique IP address. A host with multiple interfaces and hence multiple IP addresses is called a *multi-homed* host. An IP address is a 32-bit number written in the *dotted-decimal* notation, i.e., as four decimal numbers, one for each byte, separated by three periods.

The global IP address space is divided into five classes, as shown in Table 0.2. Each IP address has two parts, a *network ID*, which is common for all the IP addresses in the same network, and a *host ID*, which is unique among all hosts in the same network. Figure 0.9 shows the IP address formats for the classes, where all class A IP addresses start with "0", all

Table 0.2. *Ranges of different classes of IP addresses*

Class	From	To
A	0.0.0.0	127.255.255.255
B	128.0.0.0	191.255.255.255
C	192.0.0.0	223.255.255.255
D	224.0.0.0	239.255.255.255
E	240.0.0.0	255.255.255.255

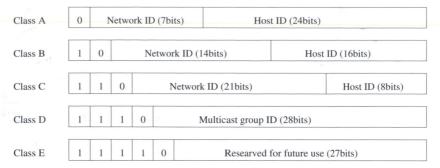

Figure 0.9. The format of IP addresses of different classes.

class B IP addresses start with "10", so on and so forth. The class of an IP address can thus be easily determined by the first number of its dotted-decimal representation. An IP address consisting of all zero bits or all one bits for the host ID field is invalid for a host IP address.

As shown in Fig. 0.9, a class A (or class B) address uses 24 bits (or 16 bits) as the host ID. Institutions assigned with a class A or B network address usually do not have that many hosts in a single network, resulting in a waste of IP addresses and inconvenience in network administration and operation. In order to provide the flexibility in network administration and operation, the *subnetting* technique was introduced, where an IP address is further divided into three levels: a network ID, a *subnet ID*, and a host ID. With subnetting, IP addresses can be assigned using a finer granularity, e.g., a small organization can be assigned a subnet address that just satisfies its requirement. In addition, with subnetting, an organization can divide its assigned network space into a number of subnets, and assign a subnet to each department. The subnets can be interconnected by routers (see Section 0.3), resulting in better performance, stronger security, and easier management.

By using Table 0.2 and Fig. 0.9, it is possible to determine the network ID of an IP address. In order to determine the subnet ID and host ID, a *subnet mask* is used to indicate how many bits are used for the host ID. A subnet mask is a 32-bit word with "1" bits for the bit positions used by the network ID and subnet ID, and "0" bits for bit positions used by the host ID. By using a subnet mask, a class A, class B or even class C network address can be subnetted based on how many subnets and how many hosts per subnet are needed.

Figure 0.10 shows how, for the same class B IP address, two different subnet masks result in two different class B arrangements. In both examples, the network ID consists of the first 16 bits since it is a class B network

	16bits	8bits	8bits
Class B	Network ID = 128.238.	Subnet ID	Host ID

Subnet Mask: 1 1 1 1 1 1 1 1 1 1 1 1 1 1 1 1 1 1 1 1 1 1 1 1 0 0 0 0 0 0 0 0
= 0xFFFFFF00 = 255.255.255.0

	16bits	10bits	6bits
Class B	Network ID = 128.238.	Subnet ID	Host ID

Subnet Mask: 1 1 1 1 1 1 1 1 1 1 1 1 1 1 1 1 1 1 1 1 1 1 1 1 1 1 0 0 0 0 0 0
= 0xFFFFFFC0 = 255.255.255.192

Figure 0.10. An example of subnet masks for two different class B subnet design.

address. The first example uses a 24-bit subnet mask, resulting in a 8-bit subnet ID and a 8-bit host ID. Therefore, there could be $2^8 = 256$ subnets and $2^8 - 2 = 254$ hosts[2] in each subnet with this subnetting scheme. In the second example, a 26-bit subnet mask is used, resulting in a 10-bit subnet ID and a 6-bit host ID. Therefore, there could be $2^{10} = 1024$ subnets and $2^6 - 2$ hosts in each subnet with this subnetting scheme. Given a network address, the administrator can flexibly trade off the number of bits needed for the subnet ID and for the host ID, to find a subnetting arrangement best suited for the administrative and operative requirements.

The network ID is often referred to as the *network-prefix*. When subnetting is used, the combination of the network ID and subnet ID is called the *extended-network-prefix*. In addition to using the IP address and network mask pair, a *slash-notation* is often used by network engineers, where an IP address is followed by a "/" and the number of 1's in the subnet mask. For example, the class B address arrangements in Fig. 0.10 can be expressed as 128.238.66.101/24 and 128.238.66.101/26, respectively.

With the combination of an IP address and a port number, a process running in a host is uniquely identified in the global Internet, since the IP address is unique in the Internet and the port number is unique within the host. The combination of an IP address and a port number is called a *socket*.

0.5.4 IP version 6

Since it was born, the Internet has been growing exponentially. Every new host computer being connected needs a unique IP address. The recent trends of *pervasive computing* that connects laptop computers, personal digital

---

[2] Host IDs are not allowed to be all 1's or all 0's.

assistants (PDA), and cell phones to the Internet, and *home networking* that connects consumer electronic devices and home appliances to the Internet require yet more IP addresses.

However, when the current version of IP (IPv4) was designed, it was never imagined that the size of the Internet would be so huge. According to [3], the 32-bit IPv4 addresses will be depleted between 2005 and 2015. Some short-term solutions have been proposed to slow down the depletion of IPv4 addresses, including the following.

- Subnetting. As discussed in the previous subsection, this technique uses network prefixes with IP addresses. Thus IP addresses can be assigned in a finer granularity than "classful" addressing, which improves the efficiency of IPv4 addressing.
- Network Address Translator (NAT). With this technique, a section of IP addresses can be reused by different private networks.

A long-term solution to the above problem is to change the engine of the Internet, i.e., introduce a new, improved version of IP. The next version of IP, IPv6, uses 128-bit addresses, which is four times the size of an IPv4 address. Theoretically, there could be 3.4×10^{38} different IPv6 addresses. Thus, IPv6 provides plenty of IP addresses for all devices that need an IP address, eliminating the need to conserve address space.

In addition to an enlarged IP address space, the IPv6 design keeps the good features of IPv4, while eliminating minor flaws and obsolete functions. Some major enhancements are listed.

- A simpler header format. IPv6 uses a 40-byte fixed length header format. Some fields in the IPv4 header that are not frequently used are removed. Options are now supported by extension headers that follow the 40-byte IPv6 header, and are used only when needed.
- Automatic configuration mechanisms. IPv6 has mechanisms that greatly simplify the network configuration of host computers. An IPv6 host can be used in a "plug-and-play" mode, i.e., without manual configuration. Network management and administration are greatly simplified.
- Security. IPv6 has extensions for authentication and privacy, including encryption of packets and authentication of the sender of packets. IPsec (Chapter 9) is an IPv6 protocol suite requirement.
- Realtime service support. IPv6 provides the *flow labeling* mechanism for realtime services. With the flow label, intermediate routers can easily identify the flow to which a packet belongs, allowing for differentiated service of packets from different flows. For example, IP datagrams

corresponding to a delay-sensitive application like a voice conversation can be served on a priority basis.

0.5.5 Medium access control address

The medium access control (MAC) address, also called the *hardware address*, is used in the link layer to uniquely identify a network interface. MAC addresses contain no location information. Since the MAC address is burned in, network interfaces can be used in *plug-and-play* mode. An IP address, on the other hand, contains information on the location of the network interface and is used to route packets to or from the interface. An IP address usually needs to be configured manually, or by the Dynamic Host Configuration Procotol (DHCP), which will be discussed in Chapter 8.

Different link layer protocols use different MAC addresses. The Ethernet MAC address is 48 bits long and is globally unique. The first 24 bits of an Ethernet address is called the *vendor component*, while the remaining 24 bits is called the *group identifier*. An Ethernet interface card vendor is assigned with a block of Ethernet addresses, starting with a unique vendor component. Each card made by the vendor has a common vendor component, followed by a different group identifier. An example MAC address, using the *hexadecimal* notation, is: 0x8:0:20:87:dd:88.

The ARP protocol is used to translate an IP address to the corresponding MAC address. We will discuss ARP in Section 2.2.4 and Ethernet addresses further in Section 7.2.1.

0.6 Multiple access

The simplest way of interconnecting two computer hosts is using a point-to-point link with a host on each end. As the number of hosts increases, this approach may be inadequate, since there needs to be a large number of links (i.e., $N(N-1)/2$) to fully connect N hosts. In this case, a *broadcast* network, where all the hosts share a common transmission medium, is more efficient.

In order to share the common medium (e.g. a cable or a wireless channel) efficiently, all hosts must follow a set of rules to access the medium. For example, at any time, there may be only one host allowed to transmit data. Otherwise, the data from two or more transmitting users may collide with each other and be corrupted. Hosts should be able to check the availability of the medium and to resolve a collision. In addition, since the total bandwidth of the medium is limited, it is desirable to share it efficiently in terms of the aggregate throughput of all the hosts. Furthermore, each host should have a fair chance to access the medium and should not be allowed to take it forever.

The sharing-rules are defined as *medium access control* (MAC) protocols. Two examples are: Carrier Sense Multiple Access/Collision Detection (CSMA/CD, used in Ethernet), and Carrier Sense Multiple Access/Collision Avoidance (CSMA/CA, used in wireless LANs). MAC protocols are implemented in the link layer. We will discuss CSMA/CD and CSMA/CA in Chapter 2.

0.7 Routing and forwarding

Various networks can be classified as *circuit-switched* networks and *packet-switched* networks. In a circuit switching network, an end-to-end circuit is set up by *circuit switches* along the path. A user communication session is guaranteed with a fixed amount of bandwidth, which is useful for many applications with *quality of service* (QoS) requirements. However, the bandwidth will be wasted if the users have no data to send, since the circuit is not shared by other users. On the other hand, the bandwidth of a network link is shared by all the users in a packet switching network. As the name suggests, user data is partitioned and stored in a sequence of packets and sent through the network. In such networks, *packet switches* route the packets, hop by hop, to the destination using information stored in the packet headers and information learned about the network topology.

Another dimension of classifying networks is defined by how the packets belonging to the same session are treated. In a *connectionless* network, every packet is self-contained, i.e., with sufficient routing information, and is treated independently, while in a *connection-oriented* network, an end-to-end connection is first set up and each packet belonging to the same session is treated consistently. Table 0.3 gives examples of how current networks fall in this classification scheme.

Routing and forwarding are the main functions of the network layer. The IP modules in the hosts and the internet routers are responsible for delivering packets from their sources to their destinations. Routing and

Table 0.3. *Classification of networks*

	Packet switching	Circuit switching
Connectionless	The Internet	–
Connection-oriented	Asynchronous Transfer Mode (ATM) networks	Plain Old Telephone Service (POTS)

forwarding consist of two closely related parts: maintaining network topology information and forwarding packets. Hosts and routers must learn the network topology in order to know where the destinations are, by exchanging information on connectivity and the quality of network links. The learned information is stored in a data structure called *routing tables* in hosts and routers. Routing tables are created or maintained either manually or by dynamic routing protocols. When there is a packet to deliver, a host or a router consults the routing table on where to route the packet. An end-to-end path consists of multiple routers. Each router relays a packet to the next-hop router which brings it closer to its destination. We will examine routing and forwarding in the Internet in Chapter 4.

0.8 Congestion control and flow control

Internet routers forward packets using the *store-and-forward* technique, i.e., an incoming packet is first stored in an input buffer, and then forwarded to the output port buffer, queued for transmission over the next link. Usually the buffer in a router is shared by many data flows belonging to different source-destination pairs. If, in a short period, a large number of packets arrive, the output port may be busy for a while and the buffer may be fully occupied by packets waiting for their turn to be forwarded (i.e., the router is *congested*). A similiar situation may occur at a destination host, which may be receiving packets from multiple sources. The packets received are first stored in a buffer, and then sent to the application processes. If the packet arriving rate is higher than the rate at which the packets are removed from the buffer, the receiving buffer may be fully occupied by packets waiting to be processed. In addition, hosts and routers are heterogeneous in terms of their processing capability and network bandwidth. In the case of a fast transmitter and a slow receiver, the receiver's buffer may get full. When the buffer,

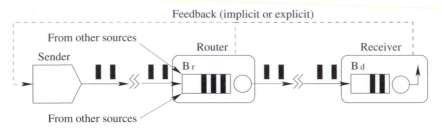

Figure 0.11. An illustration of flow control and congestion control in the Internet.

either at the receiver or at an intermediate router, is full, arriving packets have to be dropped since there is no space left to store them. Packet losses are undesirable since they degrade the quality of the communication session.

In the Internet, congestion control and flow control are used to cope with these problems. The basic idea is to let the source be adaptive to the buffer occupancies in the routers and the receiver (see Fig. 0.11, where the router has a finite buffer size B_r and the receiver has a finite buffer size B_d.). For example, the receiver may notify the sender how much data it can receive without a buffer overflow. Then the sender will not send more data than the amount allowed by the receiver. In the router case, the sender may be explicitly notified about the congestion in the router, or infer congestion from received feedback. Then the source will reduce its sending rate until the congestion is eased. TCP uses *slow start* and *congestion avoidance* to react to congestion in the routers, and to avoid receiver buffer overflow. We will discuss TCP congestion control and flow control in Chapter 6.

0.9 Error detection and control

When a packet is forwarded along its route, it may be corrupted by transmission errors. Many TCP/IP protocols use the *checksum* algorithm (or *parity check*) to detect bit errors in the header of a received packet. Suppose the checksum header field is K bits long (e.g., $K = 16$ in IP, UDP, and TCP). The value of the field is first set to 0. Then, the K-bit *one's complement sum* of the header is computed, by considering the header as a sequence of K-bit words. The K-bit one's complement of the sum is stored in the checksum field and sent to the receiver. The receiver, after receiving the packet, calculates the checksum over the header (including the checksum field) using the same algorithm. The result would be all ones if the header is error free. Otherwise, the header is corrupted and the received packet is discarded. IP, ICMP, IGMP, UDP and TCP use this algorithm to detect errors in the headers.

Ethernet, on the other hand, uses the *cyclic redundancy check* (CRC) technique to detect errors in the entire frame. With CRC, the entire frame is treated as a single number, and is divided by a predefined constant, called the CRC *generator*. The remainder of the division operation is appended to the frame (as the trailer) and sent to the receiver. After receiving the frame, the receiver performs the same division and compares the remainder with the received one. If the two are identical, there is no error in the frame. Otherwise, the frame is corrupted and should be discarded.

In addition to bit errors in a received packet, packets may be lost if there is congestion in the network, or if an incorrect route is used. *Sequence numbers* can be used to detect this type of error. With this technique, the sender and the receiver first negotiate an *initial sequence number*. Then the sender assigns a unique sequence number to each packet sent, starting from the initial sequence number and increased by one for each packet sent. The receiver can detect which packets are lost by ordering the received sequence numbers and looking for gaps in them.

When a packet loss is detected, the receiver may notify the sender, and request for a retransmission of the lost packet. In addition, the sender can use other error control schemes, such as forward error correction (FEC), in the application layer for better protection of the application data. We will examine TCP error control in Chapter 6.

0.10 Header formats of the protocols

The basic control functions discussed in the previous sections are implemented in different layers, while the information used by the control functions are carried in the packet headers. In this section, we examine the header formats of Ethernet, IP, UDP and TCP, which will be frequently used in discussions and data analysis in the following chapters.

0.10.1 Ethernet frame format

The laboratory experiments in this book are all based on Ethernet LANs. Fig. 0.12 shows the Ethernet frame format. The first 6 bytes give the `Destination Ethernet (MAC) Address`, while the next 6 bytes give the `Source Ethernet Address`. Next comes the 2-byte `Frame Type` field which is used to identify the payload of the Ethernet frame. For

Destination Address	Source Address	Frame Type	Data	CRC
6 bytes	6 bytes	2 bytes	46–1500 bytes	4 bytes

Figure 0.12. Ethernet frame format.

Version	Hdr Len	Differentiated Services		Total Length	
Identification			Flags	Fragment Offset	
Time to Live		Protocol		Header Checksum	
Source IP Address					
Destination IP Address					
Options (if any, <= 40 bytes)					
Data					

Figure 0.13. IP header format.

example, this field is set to 0x0800 for IP datagrams, 0x0806 for ARP requests and replies, and 0x0835 for RARP requests and replies. The 4-byte trailer is the CRC bits used for error control.

0.10.2 IP header format

The format of the IP header is given in Fig. 0.13. If no option is present, the size of the IP header is 20 bytes. Some of the fields are introduced below, other fields will be explained in later chapters.

- Version: 4 bits. The version of IP used, which is four for IPv4.
- Header Length: 4 bits. The header length in 32-bit words.
- Differentiated Services: 8 bits. Specifies how the upper layer protocol wants the current datagram to be handled. Six bits of this field are used as a differential service code point (DSCP) and a two-bit currently unused (CU) field is reserved.
- Total Length: 16 bits. The IP datagram length in bytes, including the IP header.
- Identification: 16 bits. Contains an integer that identifies the current datagram.
- Flags: 3 bits. Consists of a 3-bit field of which the lower two bits control *fragmentation*. The highest order bit is not used.
- Fragment Offset: 13 bits. Indicates the position of the fragment's data relative to the beginning of the data in the original datagram. It allows the destination IP process to properly reconstruct the original datagram.

Source Port Number	Destination Port Number
Length	Checksum

Figure 0.14. UDP header format.

32-bit Source IP Address		
32-bit Destination IP Address		
0x00	8-bit Protocol (0x17)	16-bit UDP Length

Figure 0.15. The pseudo-header used in UDP checksum computation.

- **Time to Live:** 8 bits. A counter that is decremented by one each time the datagram is forwarded. A datagram with 0 in this field is discarded.
- **Protocol:** 8 bits. The upper layer protocol that is the source or destination of the data. The protocol field values for several higher layer protocols are: 1 for ICMP, 2 for IGMP, 6 for TCP, and 17 for UDP.
- **Header Checksum:** 16 bits. Calculated over the IP header to verify its correctness.
- **Source IP Address:** 32 bits. The IP address of the sending host.
- **Destination IP Address:** 32 bits. The IP address of the receiving host.

0.10.3 UDP header format

The UDP header format is shown in Fig. 0.14. The Port Number fields identify sending and receiving applications (processes). Given their 16-bit length, the maximum port number is $2^{16} - 1 = 65,535$. The 16-bit Length, measured in bytes, ranges from 8 bytes (i.e., data field can be empty) to $2^{16} - 1 = 65,535$ bytes. The 16-bit Checksum is computed using the UDP header, UDP data, and a pseudo-header consisting of several IP header fields, as shown in Fig. 0.15. Using the checksum is optional and this field can be set to 0x0000 if it is not used.

0.10.4 TCP header format

The TCP header format is shown in Fig. 0.16. The fields are explained below. A more detailed discussion of TCP can be found in Chapter 6.

Source Port Number		Destination Port Number	
Sequence Number			
Acknowledgement Number			
Hdr Len.	Reserved	Flags	Window Size
TCP Checksum		Urgent Pointer	
Options (if any)			
Data (optional)			

Figure 0.16. TCP header format.

- Source Port Number: 16 bits. The port number of the source process.
- Destination Port Number: 16 bits. The port number of the process running in the destination host.
- Sequence Number: 32 bits. Identifies the byte in the stream of data from the sending TCP to the receiving TCP. It is the sequence number of the first byte of data in this segment represents.
- Acknowledgement Number: 32 bits. Contains the next sequence number that the destination host wants to receive.
- Header Length: 4 bits. The length of the header in 32-bit words.
- Reserved: 6 bits. Reserved for future use.
- Flags: There are 6 bits for flags in the TCP header, each is used as follows.
  - URG: If the first bit is set, an urgent message is being carried.
  - ACK: If the second bit is set, the acknowledgement number is valid.
  - PSH: If the third bit is set, it is a notification from the sender to the receiver that the receiver should pass all the data received to the application as soon as possible.
  - RST: If the fourth bit is set, it signals a request to reset the TCP connection.
  - SYN: The fifth bit of the flag field of the packet is set when initiating a connection.
  - FIN: The sixth bit is set to terminate a connection.
- Window Size: 16 bits. The maximum number of bytes that a receiver can accept.
- TCP Checksum: 16 bits. Covers both the TCP header and TCP data.
- Urgent Pointer: 16 bits. If the URG flag is set, the pointer points to the last byte of the urgent message in the TCP payload. More specifically, the last byte of the urgent message is identified by adding the urgent pointer value to the sequence number in the TCP header.

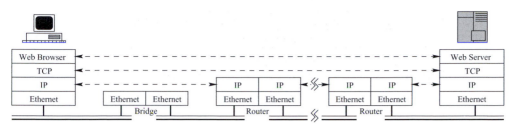

Figure 0.17. An example.

0.11 An example: how TCP/IP protocols work together

In this section, we show how a packet is forwarded from the source to the destination. As shown in Fig. 0.17, assume a user, named Bob, wants to book an air ticket from the website: `http://www.expedia.com`. Here is what happens in the system kernel and in the network.

First, Bob needs to know the domain name `www.expedia.com`, e.g., from a TV commercial or a web advertisement. If he happens to know the IP address corresponding to this domain name, he can use the IP address instead.

The remote computer with the domain name `www.expedia.com` is a *web server*, which is always running and provides the *web service*. Bob can use a web browser, which is a *web client*, to request and receive web service, i.e., to browse a web page. The HyperText Transfer Protocol (HTTP) is used by the web server and web browser. Most of the network services are provided using such a client–server architecture. We will discuss the client–server architecture in Chapter 5, and we will examine a web server in Chapter 8.

Bob starts a web browser, e.g., `Mozilla`, in his computer. Then he types `http://www.expedia.com/index.html` in the `Location` input area. The prefix `http` indicates the application layer protocol for this transaction, followed by the domain name of the web server, `www.expedia.com`, and the target file, `index.html`, in the server.

Next, the web browser needs to translate the domain name to an IP address, since domain names are not recognizable by the TCP/IP kernel. This is done via a query–response process using a protocol called the Domain Name System (DNS). The web browser invokes a function in the TCP/IP kernel called `gethostbyname()`, to send a DNS query which in essence asks "what is the IP address of '`www.expedia.com`'?" The query is sent to the host's DNS server, which is preconfigured in a file in the host, or is obtained dynamically using a protocol called Dynamic Host Configuration Protocol (DHCP) every time when the host bootstraps. A DNS server is a host maintaining a database of domain names and IP addresses. When the server receives a DNS query, it searches its database and sends a response to the querying host with the corresponding IP address. If the DNS server does not know the IP address of `www.expedia.com`, it may further query other DNS servers.

After receiving the DNS reply, the client tries to establish a *TCP connection* to the web server, since TCP is the transport layer protocol used by HTTP. The TCP/IP code is in the system kernel, but an application process can call the `socket` *application programming interface* (API) for TCP/IP services. Each application process invoking the socket API will be assigned a unique port number. The port number is carried in all the packets sent by and destined to this process. When the TCP connection is set up, the application data can be transmitted. The initial application data is a `HTTP request` message for the `index.html` file from the web server. It is sent down to the TCP layer and encapsulated in a *TCP segment*. The TCP header consists of the fields used for end-to-end flow control, congestion control, and error control, which are essential to providing an end-to-end stream-based reliable service. We will examine the use of port numbers and the concept of multiplexing in Chapter 1, study TCP in Chapter 6, and study socket API in Chapter 8.

Next, the TCP segment will be sent down to the IP layer and encapsulated in an IP *datagram*. The IP layer is responsible for forwarding the IP datagram to its destination. In order to deliver a packet to a remote host, each host or router maintains a routing table storing routing information. Only the next-hop IP address to a destination is stored. When a host has an IP datagram to sent, or when a router receives a datagram to forward, it searches the routing table to find the next-hop router, and forwards the datagram to that router. The routing table can be set manually, or dynamically by routing protocols. We will examine IP routing and configure a commercial router in Chapter 4.

In this example, the IP module of Bob's host finds the next-hop router in its routing table, and sends the IP datagram and the next-hop router's IP address down to the MAC layer. This host uses an Ethernet card and the IP datagram is further encapsulated within an Ethernet frame. The Ethernet driver is responsible for delivering the Ethernet frame to the interface of the next-hop router. Before sending the Ethernet frame out, the device driver has to resolve the next-hop IP address, since it only recognizes Ethernet MAC addresses. An ARP request is broadcast, querying the MAC address associated with the target IP address. When the router interface receives this ARP request, it responses with an ARP reply containing its MAC address. Then, the frame is sent on the medium after the ARP reply is received and the destination MAC address is learned. Note that whenever the host sends a frame, it uses the CSMA/CD multiple access algorithm to access the channel and may *backoff* if collision occurs. We will examine the operation and configuration of an Ethernet interface in Chapter 2.

Bob's local network consists of several LAN segments. Several IEEE 802.1d bridges, which are *self-configuring* and *transparent*, are used to connect the LAN segments. The *spanning tree* algorithm is running in the bridges to avoid loops in the local network. In this example, the Ethernet frame is first transmitted on the host's LAN segment, and then forwarded to the router interface by an intermediate bridge. We will examine bridges and the spanning tree protocol in Chapter 3.

Subsequently, the IP datagram is forwarded hop-by-hop by the intermediate routers along the route towards its destination. Some of the routers may be connected by point-to-point long-haul connections running the SDH/SONET protocol. Finally, the remote host's MAC module receives the Ethernet frame. The packet is delivered to the upper layers. At each layer, the corresponding header is stripped and examined. The information carried in the headers is used for such functions such as routing and forwarding, error control, flow control, and congestion control. In addition, the information is also used to identify which higher layer module the payload data belong to. When the Web server at the application layer receives the `HTTP request` message, it assembles an `HTTP response` message containing the requested file, and sends the response to the client. The response message is forwarded back to Bob's host, through a similar procedure. Finally, Bob can see the homepage of `www.expedia.com` in his web browser.

1 Linux and TCP/IP networking

The Linux philosophy is 'Laugh in the face of danger'. Oops. Wrong One. 'Do it yourself'. Yes, that's it.
Linus Torvalds

1.1 Objectives

- Getting acquainted with the lab environment.
- Getting acquainted with the Linux operating system.
- Preview of some TCP/IP diagnostic tools.
- Capturing and analyzing the link layer, IP, and TCP headers.
- Understanding the concept of encapsulation.
- Understanding the concept of multiplexing using port numbers, the IP *protocol* field, and the Ethernet *frame type* field.
- Understanding the client–server architecture.

1.2 Linux and TCP/IP Implementations

1.2.1 TCP/IP Implementations

The TCP/IP protocol architecture was first proposed in the Cerf and Kahn paper [1]. Since then, the TCP/IP protocol family has evolved over time into a number of different versions and implementations. The first widely available release of TCP/IP implementation is the 4.2 Berkeley Software Distribution (BSD) from the Computer Systems Research Group at the University of California at Berkeley. Many implementations of TCP/IP protocols are based on the public domain BSD source code, both for Unix and non-Unix systems, as well as public domain implementations and implementations from various vendors.

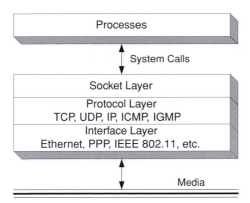

Figure 1.1. Organization of the networking code.

Solaris and FreeBSD are two examples of Unix TCP/IP implementations. Solaris is an operating system developed by Sun Microsystems. It supports both the SPARC platform and the x86 platform. FreeBSD is a Unix operating system derived from BSD. It was developed and is maintained by a large team of individuals. FreeBSD also supports multiple platforms and is available free of charge. Linux is a popular Unix-type operating system. It was originally created by Linus Torvalds and further improved by developers all over the world. Linux is developed under the *GNU General Public License*. The Linux source code is available in the public domain and the system kernel is recompilable. Linux can also be embedded in small devices, such as cellphones and PDAs. These features make Linux very popular in the computer and networking research communities. In addition, Linux is gaining support from major computer vendors, such as IBM, Oracle, and Dell.

From an implementation point of view, the networking code can be organized into four layers, as illustrated in Fig. 1.1. Most applications are implemented as *user space* processes, while protocols in the lower three layers (i.e., the transport layer, network layer, and data link layer) are implemented in the *system kernel*.[1] A user space process can obtain services provided by the kernel by invoking *system calls*. In the system kernel, the networking code is organized into three layers, namely the *socket layer*, the *protocol layer*, and the *interface layer*. The socket layer

---

[1] The core of an operating system, implementing critical system functions, e.g., managing memory and file systems, loading and executing other programs, and scheduling processes.

Table 1.1. *A few lines in the* /etc/services *file*

. . .

ftp stream tcp nowait root /usr/sbin/tcpd in.ftpd ftpd
telnet stream tcp nowait root /usr/sbin/tcpd in.telnetd

. . .

#finger stream tcp nowait root /usr/sbin/tcpd in.fingerd

. . .

is protocol independent. It provides a common interface to the user processes and hides the protocol specific details from them. The protocol layer contains the implementation of TCP/IP protocols, while the interface layer consists of *device drivers* which communicate with the network devices [4].

1.2.2 Network daemons and services

A daemon is a process running in the background of the system. Many TCP/IP services (e.g., Telnet) are handled by a daemon called **inetd**. Rather than running several network-related daemons, the **inetd** daemon works as a dispatcher and starts the necessary server processes when requests arrive. When a client wants a particular service from a remote server, the client contacts the **inetd** daemon through the server's well-known port number, which prompts **inetd** to start the corresponding server process.

The network daemons managed by **inetd** are specified in a configuration file called /etc/inetd.conf. Each service has a line in the file defining the network daemon that provides the service and its configuration parameters. Table 1.1 shows three lines in the /etc/inetd.conf file, which correspond to Ftp, Telnet, and Finger[2] services. One can comment a line, i.e., insert a # at the beginning of the line, to disable the corresponding service. For example, the Finger service in the following example is disabled. Note that there are some stand-alone network daemons that are not managed by **inetd**. For example, web service is provided by the **httpd** daemon, and DNS service is provided by the **named** daemon.

---

[2] Used to display information about a user.

In Red Hat Linux 9, **xinetd** replaces **inetd**, adding stronger security and more functionality. **xinetd** uses a simple common configuration file /etc/xinetd.conf. In addition, each service managed by **xinetd** uses an individual configuration file in the /etc/xinetd.d directory. The following is the Echo service configuration file /etc/xinetd.d/echo. It can be seen that the Echo service is enabled and uses TCP in the transport layer.

```
# default: off
# description: An echo server. This is the tcp \
# version.
service echo
{
        disable = no
        type = INTERNAL
        id = echo-stream
        socket_type = stream
        protocol = tcp
        user = root
        wait = no
}
```

Well-known port numbers are defined in the /etc/services file. A server can handle multiple clients for a service at the same time through the same well-known port number, while a client uses an ephemeral port number. The uniqueness of a communication session between two hosts is preserved by means of the port number and IP address pairs of the server and client hosts.

1.2.3 Network configurations files

When a host is configured to boot locally, certain TCP/IP configuration parameters are stored in appropriate local disk files. When the system boots up, these parameters are read from the files and used to configure the daemons and the network interfaces. A parameter may be changed by editing the corresponding configuration file.

In addition to /etc/services and /etc/inetd.conf discussed above, we now list other network configuration files.

`/etc/hosts`	Stores the host name of this machine and other machines.
`/etc/sysconfig/network`	Stores the host name and the default gateway IP address.
`/etc/sysconfig/network-` `scripts/ifcfg-eth0`	Stores the IP address of the first Ethernet interface.
`/etc/default-route`	Stores a default gateway, i.e., the IP address or the domain name of the default router.
`/etc/resolv.conf`	Stores the IP addresses of the DNS servers.
`/etc/nsswitch.conf`	Configures the means by which host names are resolved.

Solaris uses the following network configuration files stored in the `/etc` directory.

`nodename`	Host name of the machine.
`hostname.interface`	Interface IP address or the interface name.
`inet/hosts`	Stores IP addresses of the interfaces of the machine, the corresponding host name for each interface, IP addresses of the file server, and IP address and name of the default router.
`defaultdomain`	The host's fully qualified domain name.
`defaultrouter`	The name for the network interface that functions as this host's default router.
`inet/netmasks`	The network ID and the netmask if the network is subnetted.
`inet/networks`	Associates network names with network numbers, enabling applications to use and display names rather than numbers.
`nsswitch.conf`	Specifies name service to use for a particular machine.

1.3 Linux commands and tools

1.3.1 Basic Linux commands

The basic Linux commands are summarized below. See the manual pages for a list of options for each command.

- **man** *command_name*: Gets online help for *command_name*.
- **passwd**: Sets (changes) the password.
- **pwd**   : Displays the current working directory.
- **ls**   : Lists the contents of a directory.
- **more** *file_name*: Scrolls through a file.
  - To list the next page, press the space bar.
  - To go backwards, press b.
  - To quit from **more**, press q.
- **mv** *old_file_name new_file_name*: Renames a file.
  mv *file_name directory_name*: Moves a file to a directory.
  mv *old_directory_name new_directory_name*: Renames a directory.
- **rm** *file_name*: Deletes(removes) a file.
- **mkdir** *directory_name*: Creates a directory.
- **rmdir** *directory_name*: Removes a directory.
- **cd** *directory_name*: Changes the current working directory to *directory_name*. If *directory_name* is omitted, the shell is moved to your home directory.
- **cp** *file_name new_file_name*: Copies a file.
  cp *file_name directory_name*: Copies a file into *directory_name*.
- **chmod** *who op-code permission file_or_directory_name*: Changes the file access permissions.
  who: **u** user, **g** group, **o** other users, **a** all;
  op-code: + add permission, − remove permission;
  permission: **r** read, **w** write, **x** execute.
- **ps**: Process status report.
- **kill** *PID*: Terminates the process with a process ID *PID*.
- **Ctrl-c** : Terminates a command before it is finished.
- **cmp** *file1 file2*: Compares *file1* and *file2* byte by byte.
- **grep** *keyword file(s)*: Search the file(s) and outputs the lines containing the keyword.

Most of the above commands accept input from the system's *standard input* device (e.g., the keyboard) and send an output to the system's *standard output* device (e.g., the screen). Sometimes it is convenient to direct the output to another process as input for further processing, or to a file for storage. The *redirect* operator ">" directs the output to a file, as:

command > *file_name*.

With the *pipe* operator "|", two commands can be concatenated as:

command1 | command2,

where the output of **command1** is redirected as the input of **command2**.

1.3.2 Text editor

The vi Editor

The **vi** editor is one of the most popular text editors. It is the default text editor of most Linux and Unix systems.

To start **vi**, enter **vi** *file_name* at the command line. If no such file exists yet, it will be created. **vi** can be in one of the two modes, the *command mode* and the *text entry mode*. The command mode allows a user to use a number of commands to modify text. Text is inserted and modified in the text entry mode. Initially, **vi** enters the command mode and awaits instructions. To enter text, switch to the text entry mode by typing one of the following keys.

i: Text is inserted to the left of the cursor.
a: Text is appended after the cursor.
o: Text is added after the current line.
O: Text is added before the current line.

To switch back to the command mode, press the Esc key.

In the command mode, the user may use **vi**'s several editing features, such as cursor movement, text deletion, text replacement, and search operation. Some of the basic features are listed here.

h: Moves the cursor one space to the left.
j: Moves the cursor down one line.
k: Moves the cursor up one line.
l: Moves the cursor one space to the right.
Ctrl-**f**:   Scrolls down one full screen.
Ctrl-**b**:   Scrolls up one full screen.

Ctrl-**d**:   Scrolls down a half page.

Ctrl-**u**:   Scrolls up a half page.

To delete text, place the cursor over the target position, and type the following commands.

x: Delete the next single character.

dw: Delete the current word.

dd: Delete the current line.

To search for a special string, e.g. *foo*, in the text file, type */foo* in the command mode. The cursor will jump to the nearest matching position in the file. To repeat the last search, type **n** in the command mode.

To save the file and quit **vi**, press Esc (even if in the command mode, it doesn't hurt), and type :wq. To quit **vi** without saving changes to the file, use the command :q!.

Other text editors

In addition to **vi**, Red Hat Linux provides a number of graphical text editors that are intuitive and easy to use. They are especially convenient for users who are used to a windows-based interface. Examples of such graphical text editors are given below.

Emacs	A free text editor preinstalled in most Linux operating systems. It can be invoked by running **emacs**, or by choosing the system menu item: Programming/Emacs. The system menu pops up when the Red Hat icon at the lower-left corner of the workspace is clicked.
gedit	An analog to Notepad in Microsoft Windows. It can be invoked by running **gedit**, or by choosing the sytem menu item: Accessories/Text Editor.
OpenOffice.org	An analog to the Microsoft Office package. The Writer component of OpenOffice.org provides functions similar to that offered by Microsoft Word. A file created by OpenOffice.org Writer can be ported to the Microsoft Word format.

There is a graphical text editor available with the Solaris OpenWindows or Common Desktop Environment (CDE). It can be invoked by running **/usr/openwin/bin/textedit**, or from the system program menu. The system program menu can be found by clicking the right mouse-key on the background of the workspace. A new texteditor is started when the `Text Editor` menu item is chosen from the menu.

1.3.3 Window dump

The windows on the screen can be dumped into a graphic file. To dump the entire desktop area, you can simply type the `PrintScreen` key.[3] To dump a specific window (e.g., a command console), first click in the window and then type `Alt-PrintScreen`. The window is then dumped and the user is prompted for a file name to store it. The window dump is saved in the Portable Network Graphics (PNG) format. It can be opened, edited, and converted to other formats (e.g., PostScript, JPEG, or GIF) using the GIMP graphic editor supplied with Red Hat Linux.

In Solaris, a user may use **xwd** and an additional tool called **xpr** to dump a window. The pipe operator is used to conveniently redirect the output of **xwd** as the input of **xpr**, as:

> **xwd | xpr -device ps -output** *file_name*

After the shape of the mouse pointer changes, click the mouse in the target window. The keyboard bell rings once at the beginning of the dump and twice when the dump is completed. The dumped file may be examined with **Image tools** found in the **Programs** menu.

1.3.4 Using floppy disks

To format an MS-DOS formatted disk, insert the disk into the floppy disk drive and execute: **fdformat /dev/fd0**. To use an MS-DOS formatted floppy disk, insert the disk and use the **mount /dev/fd0**[4] command to mount the floppy disk to the `/mnt/floppy` directory. A new icon called "floppy" will

---

[3] Different keyboards may have different names for this key. For example, `Prnt Scrn` or `PrtSc`.
[4] `/dev/fd0` is the device name of the floppy driver. The device names and their corresponding mount points are defined in the `/etc/fstab` file. For example, the CDROM has a device name `/dev/cdrom` and is mounted to the `/mnt/cdrom` directory.

appear on the desktop after the floppy is mounted. A double-click on this icon will start a file manager which opens the floppy disk.

To manipulate the files in the floppy disk, use the same Linux commands as usual. For example, to copy a file into the floppy disk, use

> **cp** *file_name* `/mnt/floppy`.

To delete a file from the floppy disk, use

> **rm** `/mnt/floppy/`*file_name*.

The floppy disk can be unmounted by the **umount /dev/fd0** command. Then the "floppy" icon disappears. The floppy disk can be ejected manually. Note that all the opened files in the floppy disk must be closed. Also, the current directory of any of the command consoles should not be `/mnt/floppy`. Otherwise, the **umount** command will fail with a "device is busy" warning.

In Solaris, the CDROM and floppy drives are controlled by the volume manager, which is a daemon process named **vold**. When a floppy disk is inserted in the drive, **vold** does not automatically recognize it (However, it recognizes a CD automatically). The user should type the **volcheck** command, which mounts the floppy disk under the `/floppy/floppy0` directory. To eject the floppy disk, use the **eject** command. To format a MD-DOS disk, use **fdformat -v -U -d**.

1.4 Diagnostic tools

Diagnostic tools are used to identify problems in the network, as well as to help understand the behavior of network protocols. We will use the following tools extensively in the experiments.

1.4.1 Tcpdump

Tcpdump is a network traffic sniffer built on the packet capture library `libpcap`.[5] While started, it captures and displays packets on the LAN segment. By analyzing the traffic flows and the packet header fields, a

---

[5] A public domain packet capture library written by the Network Research Group (NRG) of the Information and Computing Sciences Division (ICSD) of the Lawrence Berkeley National Laboratory (LBNL) in Berkeley, California.

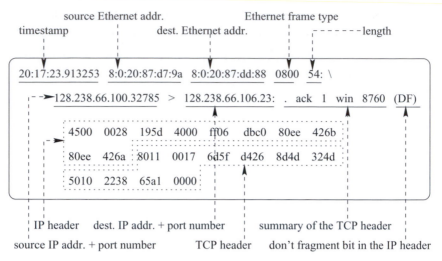

Figure 1.2. A typical tcpdump output.

great deal of information can be gained about the behavior of the protocols and their operation within the network. Problems in the network can also be identified. A packet filter can be defined in the command line with different options to obtain a desired output.

A typical output of **tcpdump** running on a 128.238.66.0 subnet is shown in Fig. 1.2. The first line of the output gives a summary of the link/IP/TCP headers, while the following data block contains the raw bits of the IP datagram.

1.4.2 Ethereal

Ethereal is a network protocol analyzer built on the packet capture library pcap. In addition to capturing network packets as in Tcpdump, Ethereal provides a user friendly graphical interface, and supports additional application layer protocols. Ethereal can also import pre-captured data files from other network monitoring tools, such as Tcpdump and Sniffer.

In the following experiments, we use Ethereal to analyze a packet trace captured by Tcpdump, since generally Ethereal does not allow a normal user to capture packets (see Section A.5).

1.5 Exercises with Linux commands

We start with a simple single segment network, where all eight computers are connected in one Ethernet segment (see Fig. 1.3). The host IP

Table 1.2. *The IP addresses of the hosts in Fig. 1.3*

Host	IP Address	Subnet Mask
shakti	128.238.66.100	255.255.255.0
vayu	128.238.66.101	255.255.255.0
agni	128.238.66.102	255.255.255.0
apah	128.238.66.103	255.255.255.0
yachi	128.238.66.104	255.255.255.0
fenchi	128.238.66.105	255.255.255.0
kenchi	128.238.66.106	255.255.255.0
guchi	128.238.66.107	255.255.255.0

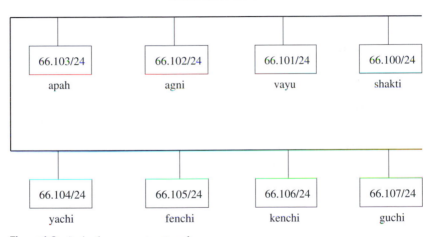

Figure 1.3. A single segment network.

addresses are given in Table 1.2. Note that the *slash-notation* is used, where "128.238.66.100/24" means an IP address of "128.238.66.100" with a subnetmask of "255.255.255.0".

Exercise 1   Login to the system. The login ID is **guest**, and the login password is **guest1**. Get acquainted with the Gnome environment, the Linux commands, text editors, and the man pages.

Exercise 2   After logging in, open a command window if one is not opened automatically, by clicking the right mouse-key on the background and choosing the New Terminal item in the menu.

In Solaris OpenWindows environment, click the right mouse key to invoke the workspace menu. Then choose `Programs/Command Tool ...` to bring up a new command window. In Solaris CDE, click the right mouse key to get the Workspace Menu. Then choose `Tools/Terminal` to bring up a new command window.

Show your login ID by typing **whoami** in the console.

Create a directory of your own, using **mkdir** *name_of_your_directory*. Change to your directory, using **cd** *name_of_your_directory*. You can save your data files for all your laboratory experiments here.

Open another command window. Run **pwd** in this and the previously opened consoles. Save the outputs in both consoles.

LAB REPORT   What is the default directory when you open a new command window? What is your working directory?

Exercise 3   Run **ps -e** to list the processes running in your host. After starting a new process by running **telnet** in another command window, execute **ps -e** again in a third window to see if there is any change in its output.

Find the process id of the **telnet** process you started, by:

ps -e | grep telnet.

Then use **kill** *process_id_of_telnet* to terminate the **telnet** process.

LAB REPORT   Is the Internet service daemon, **xinetd**, started in your system? Is **inetd** started in your system? Why?

Exercise 4   Display the file `/etc/services` on your screen, using:

more /etc/services.

Then in another console, use the redirect operator to redirect the **more** output to a file using **more /etc/services** > **ser_more**. Compare the file `ser_more` with the original **more** output in the other command window.

Copy `/etc/services` file to a local file named `ser_cp` in your working directory, using **cp /etc/services ser_cp**. Compare files `ser_more` and `ser_cp`, using **cmp ser_more ser_cp**. Are these two files identical?

Concatenate these two files using **cat ser_more ser_cp** > **ser_cat**.

Display the file sizes using **ls -l ser***. Save the output. What are the sizes of files `ser_more`, `ser_cp`, and `ser_cat`?

LAB REPORT   Submit the **ls** output you saved in this exercise and answer the above questions.

1.6 Exercises with diagnostic tools

Exercise 5   Read the **man** pages for the following programs:

arp	arping	ifconfig	tcpdump
ping	netstat	route	ethereal

The **arping** command is not provided in Solaris 8.0.

Study the different options associated with each command. Throughout this lab you will use these commands rather extensively.

LAB REPORT   Explain the above commands briefly. Two or three sentences per command would be adequate.

Exercise 6   In this exercise, we will use **tcpdump** to capture a packet containing the link, IP, and TCP headers and use **ethereal** to analyze this packet.

First, run **tcpdump -enx -w exe6.out**. You will not see any **tcpdump** output, since the **-w** option is used to write the output to the exe6.out file.

Then, you may want to run **telnet** *remote_host*[6] to generate some TCP traffic. After you login the remote machine, terminate the **telnet** session and terminate the **tcpdump** program.

Next, you will use **ethereal** to open the packet trace captured by **tcpdump** and analyze the captured packets. To do this, run **ethereal -r exe6.out &**. The **ethereal** Graphical User Interface (GUI) will pop up and the packets captured by **tcpdump** will be displayed.

For your report, you need to save any one of the packets that contain the link, IP, and TCP headers. Carry out the following instructions.

1. Click on a TCP packet from the list of captured packets in the **ethereal** window. Then go to the Edit menu and choose Mark Frame.
2. Go to the File menu and choose Print. In the Ethereal:Print dialog that pops up, check File, Plain Text, Expand all levels, Print detail,

---

[6] We use *remote_host* to denote the IP address of a remote host, i.e., a machine other than the one you are using.

and Suppress unmarked frames. Then, enter the output text file name, e.g., headers.txt, and click the OK button. The marked packet is now dumped into the text file, with a detailed list of the name and value of every field in all the three headers.

LAB REPORT   Draw the format of the packet you saved, including the link, IP, and TCP headers (See Figs 0.12, 0.13, and 0.16 in Chapter 0 of this guide), and identify the value of each field in these headers. Express the values in the decimal format.

LAB REPORT   What is the value of the protocol field in the IP header of the packet you saved? What is the use of the protocol field?

Exercise 7   In a manner similar to the previous exercise, we will run **tcpdump** to capture an ARP request and an ARP reply,[7] and then use **ethereal** to analyze the frames.

Run **tcpdump -enx -w exe7.out** to capture all the packets on the LAN segment.

If there is no arp requests and replies in the network, generate some using **arping** *remote_machine*.

When Solaris 8.0 is used, you can generate an ARP request and an ARP reply by running **telnet** to a remote machine. Note this remote machine should be a different machine from the one you used in Exercise 6.

After you see several ARP replies in the **arping** output, terminate the **arping** and the **tcpdump** program. Open the **tcpdump** trace using **ethereal -r exe7.out &**. Print one ARP request and one ARP reply using **ethereal**.

LAB REPORT   What is the value of the frame type field in an Ethernet frame carrying an ARP request and in an Ethernet frame carrying an ARP reply, respectively?

What is the value of the frame type field in an Ethernet frame carrying an IP datagram captured in the previous exercise?

What is the use of the frame type field?

Exercise 8   Using the **tcpdump** utility, capture any packet on the LAN and see the output format for different command-line options. Study the various expressions for selecting which packets to be dumped.

---

[7] We will examine the Address Resolution Protocol (ARP) in the next Chapter. For this exercise, the purpose is to examine the use of the frame type field in an Ethernet frame.

For this experiment, use the **man** page for **tcpdump** to find out the options and expressions that can be used.

If there is no traffic on the network, you may generate traffic with some applications (e.g. **telnet**, **ping**, etc.).

LAB REPORT   Explain briefly the purposes of the following **tcpdump** expressions.

> **tcpdump udp port 520**
> **tcpdump -x -s 120 ip proto 89**
> **tcpdump -x -s 70 host** *ip_addr1* **and** (*ip_addr2* **or** *ip_addr3*)
> **tcpdump -x -s 70 host** *ip_addr1* **and not** *ip_addr2*

1.7 Exercises on port numbers

Exercise 9   Start **tcpdump** in a command window to capture packets between your machine and a remote host using:

> **tcpdump -n -nn host** *your_host* **and** *remote_host*[8].

Execute a TCP utility, **telnet** for example, in another command window.

When you see a TCP packet in the **tcpdump** output, terminate **tcpdump** and save its output.

LAB REPORT   What are the port numbers used by the remote and the local computer? Which machine's port number matches the port number listed for **telnet** in the /etc/services file?

Exercise 10   Start **tcpdump** in one command window using:

> **tcpdump -n -nn host** *your_host* **and** *remote_host*.

Then, **telnet** to the remote host from a second command window by typing **telnet** *remote_host*. Again issue the same **telnet** *remote_host* command from a third command window. Now you are opening two **telnet** sessions to the same remote host simultaneously, from two different command windows.

Check the port numbers being used on both sides of the two connections from the output in the **tcpdump** window. Save a TCP packet from each of the connections.

LAB REPORT   When you have two **telnet** sessions with your machine, what port number is used on the remote machine?

Are both sessions connected to the same port number on the remote machine?

---

[8] For some older versions of **tcpdump**, the −n −nn options is combined into one single −n option.

What port numbers are used in your machine for the first and second **telnet**, respectively?

LAB REPORT   What is the range of Internet-wide well-known port numbers? What is the range of well-known port numbers for Unix/Linux specific service? What is the range for a client port number? Compare your answer to the well-known port numbers defined in the /etc/services file. Are they consistent?

LAB REPORT   Explain briefly what a socket is.

2   A single segment network

Metcalfe's Law: "The value of a network grows as the square of the number of its users."

Robert Metcalfe

2.1 Objectives

- Network interfaces and interface configuration.
- Network load and statistics.
- The Address Resolution Protocol and its operations.
- ICMP messages and Ping.
- Concept of subnetting.
- Duplicate IP addresses and incorrect subnet masks.

2.2 Local area networks

Generally there are two types of networks: point-to-point networks or broadcast networks. A point-to-point network consists of two end hosts connected by a link, whereas in a broadcast network, a number of stations share a common transmission medium. Usually, a point-to-point network is used for long-distance connections, e.g., dialup connections and SONET/SDH links. Local area networks are almost all broadcast networks, e.g., Ethernet or wireless local area networks (LANs).

2.2.1 Point-to-Point networks

The Point-to-Point Protocol (PPP) is a data link protocol for PPP LANs. The main purpose of PPP is encapsulation and transmission of IP datagrams,

Flag	Addr	Ctrl	Protocol	Data	CRC	Flag
7E	FF	03				7E
1 byte	1 byte	1 byte	2 bytes	<=1500 bytes	2 bytes	1 byte

Protocol	Data
0021	IP Datagram
C021	Link Control Data
8021	Network Control Data

Figure 2.1. PPP frame format.

or other network layer protocol data, over a serial link. Currently, most dial-up Internet access services are provided using PPP.

PPP consists of two types of protocols. The Link Control Protocol (LCP) of PPP is responsible for establishing, configuring, and negotiating the data-link connection, while for each network layer protocol supported by PPP, there is a Network Control Protocol (NCP). For example, the IP Control Protocol (IPCP) is used for transmitting IP datagrams over a PPP link. Once the link is successfully established, the network layer data, i.e., IP datagrams, are encapsulate in PPP frames and transmitted over the serial link.

The PPP frame format is shown in Fig. 2.1. The two `Flag` fields mark the beginning and end points of a PPP frame. The `Protocol` field is used to multiplex different protocol data in the same PPP frame format. Since there are only two end hosts in a PPP LAN, neither an addressing scheme nor medium access control are needed.

2.2.2 Ethernet LANs

In a broadcast network where a number of hosts share a transmission medium, a set of rules, or protocols, are needed in order to resolve collisions and share the medium fairly and efficiently. Such protocols are called medium access control (MAC) protocols. Examples of MAC protocols proposed for various networks are: Aloha, Carrier Sense Multiple Access/Collision Detection (CSMA/CD), and Carrier Sense Multiple Access/Collision Avoidance (CSMA/CA).

Ethernet has been an industry standard since 1982 and is based on the first implementation of CSMA/CD by Xerox. In an Ethernet LAN, all of the hosts are connected to a common channel. When a host has a frame to

Table 2.1. *The exponential backoff algorithm used in CSMA/CD*

1. Set a "time slot" to $2a$.
2. After the i-th collision, the random transmission time is uniformly chosen from a range of 0 to $2^i - 1$ time slots.
3. Do not increase the random time range if $i \geq 10$.
4. Give up after 16 collisions and drop the frame.

send, it first senses the channel to see if there is any transmission going on. If the channel is busy, the host will wait until the channel becomes idle. Otherwise, the host begins transmission if the channel is idle. Assume the maximum end-to-end propagation delay is a seconds. After the first bit is transmitted, the host keeps on sensing the channel for $2a$ seconds. If there is no collision detected during this period, the entire frame is assumed to be transmitted successfully. This is because it takes at most a seconds for all the hosts to hear this transmission, and another a seconds to hear any possible collision with another transmission. When a collision is detected, all hosts involved in the collision stop transmitting data and start to *backoff*, i.e., wait a random amount of time before attempting to transmit again. The random time is determined by the *exponential backoff* algorithm given in Table 2.1.

In addition to attaching all hosts to a common cable or a hub, an Ethernet LAN can be built using *Ethernet switches* with a star topology. Ethernet switches, also called *switched hubs*, are MAC layer devices that switch frames between different ports. An Ethernet switch offers guaranteed bandwidth for the LAN segments connected to each port and separates a LAN into collision domains. If each Ethernet switch port is connected to a single host only, CSMA/CD operation is not required. However, in order for the switch to deal with traffic congestion, the switch may generates a false collision signal (backpressure) to make the transmitting host back off.

2.2.3 IEEE 802.11 wireless LANs

In addition to PPP and Ethernet LANs, wireless LANs (WLANs) using the IEEE 802.11 protocols have rapidly gained popularity in recent years. In a WLAN, computers share a wireless channel. Thus there is no need to install cables, and the computers can be mobile.

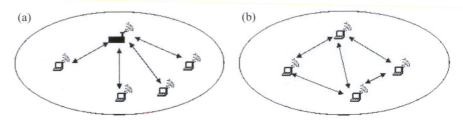

Figure 2.2. Different operation modes of IEEE 802.11 WLANs. (a) The infrastructure mode. (b) The ad-hoc mode.

An IEEE 802.11 WLAN can be configured to work in two modes: the *infrastructure* mode and the *ad-hoc* mode. In the infrastructure mode shown in Fig. 2.2(a), fixed *access points* are used. These access points are connected to the wireline network. Each access point communicates with hosts within its transmission range and serves as a gateway for the hosts. When an active mobile host moves from one access point to another, *handoff* techniques can be applied to switch the connection from the original access point to the new access point without an interruption. In addition, multiple access points can be configured to work together to provide extended coverage. In the ad hoc mode shown in Fig. 2.2(b), there is no need for access points. Host computers can communicate with each other as long as they are in each other's transmission range.

In WLANs, CSMA/CD is inadequate because collision detection cannot be performed effectively in a wireless channel. Rather, CSMA/CA is used for medium access control. In CSMA/CA, a host first senses the medium when it has a frame to send. If the medium remains free for a certain period of time (called the Distributed Coordination Function (DCF) Inter-Frame Space (DIFS)), the host begins transmitting data. When the transmission is over, it waits for an acknowledgement from the receiving host. If no acknowledgement received, it assumes that a collision occurred and prepares to retransmit. On the other hand, if the medium is busy, the host waits for the end of the current frame transmission plus a DIFS, and then begins a backoff procedure like in the case of CSMA/CD protocol. Backoff is performed as follows. The host first chooses a random number within a certain range as a backoff time, and then listens to the wireless channel to determine if it is free or busy. The backoff time is decremented by one if the medium is free in a time slot. However, the host stops decrementing the backoff time if the medium is busy during a time slot, and resumes decrementing it only when the medium becomes free again. When the backoff time becomes 0 and the channel is idle, the host attemps a transmission.

(a)   (b)

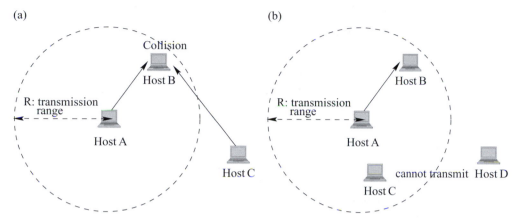

Figure 2.3. The hidden terminal and exposed terminal problems of IEEE 802.11 WLANs. (a) The hidden terminal problem. (b) The exposed terminal problem.

The frame will be dropped if the maximum number of retransmissions is reached.

WLAN has the *hidden terminal* and *exposed terminal* problems inherent from the use of wireless channels. Consider the scenario shown in Fig. 2.3(a), where Host A and C are far away and cannot hear each other. Host B is somewhere in between and can hear both Host A and C. Host A is transmitting data to Host B, and Host C also has data to send to B. If CSMA/CA is used, Host C senses an idle channel because it cannot hear Host A's transmission. Host C therefore begins transmitting data to Host B and collision occurs at Host B. Neither Host A nor C can detect the collision in this case. This is called the hidden terminal problem. Now let us consider a different scenario, as shown in Fig. 2.3(b). There are four hosts in the system. Host A and D are far from each other, and Host B and C are in between. Host A is transmitting data to Host B, while Host C has data for Host D. If Host D is out of the tranmission range of Host A and Host B is out of the transmission range of Host C, Host C can start transmitting without causing or collision at Host B and D. However, if CSMA/CA is used, Host C detects a busy channel and will wait till the current transmission is over, resulting in a waste of bandwidth.

In WLAN, the hidden terminal problem is solved by sending request-to-send (RTS) and clear-to-send (CTS) messages before the data transmission. When a host wants to send a data frame, it first sends a RTS carrying the time needed to transmit the frame. The receiving host, if it is free, responds with a CTS. All other hosts that hear the RTS or CTS will mark the medium as busy for the duration of the requested transmission. In the above example, Host

C first sends a RTS to Host B. Since Host B is engaged in the transmission from Host A, it will not return a CTS. Host C cannot transmit without a CTS, and the collision is avoided. However, the exposed terminal problem illustrated in Fig. 2.3(b) is not solved by this mechanism.

IEEE 802.11 is also popularly known as *Wi-Fi* (Wireless Fidelity). The Wi-Fi Alliance, a nonprofit international association, certifies interoperability of WLAN products based on the IEEE 802.11 standard (http://www.wi-fi.org).

2.2.4 The Address Resolution Protocol

ARP and RARP

As shown in Fig. 1.1, the protocol layer uses the service provided by the interface layer to send and receive IP datagrams. However, IP addresses used in the protocol layer are not recognizable in the interface layer where physical addresses (or MAC addresses) are used. Furthermore, different kinds of physical networks use different addressing schemes. In order to run TCP/IP over different kinds of physical transmission media, the link layer provides the function that maps an IP address to a physical network address. The protocol that performs this translation is the address resolution protocol (ARP). When a mapping from MAC address to IP address is needed, the reverse address resolution protocol (RARP) is used. Since each type of physical network has different protocol details, there are different ARP RFCs for Ethernet, Fiber-Distributed Data Interface (FDDI), Asynchronos Transfer Mode (ATM), Fiber Channel, and other types of physical networks. In this section, we focus on Ethernet ARP.

When the device driver receives an IP datagram from the IP layer, it first broadcasts an *ARP request* asking for the MAC address corresponding to the destination IP address. After receiving the ARP request, the destination host (with the target IP address) will return an *ARP reply*, telling the sender its MAC address. After this *question-and-answer* process, the source device driver can assemble an Ethernet frame, with the received MAC address as destination MAC address and with the IP datagram as the payload, and then transmit it on the physical medium. Obviously, it is inefficient to have an ARP request/reply exchange for each IP datagram that is sent. Instead, each host maintains an ARP cache, which contains recently resolved IP addresses. When a host has an IP datagram to send to another host, it first checks its ARP cache. If an entry for the destination IP is found, the

Hardware Type	Protocol Type	Hardware Size	Protocol Size	Operation Field	Sender Eth. Addr.	Sender IP Addr.	Target Eth. Addr.	Target IP Addr.
2	2	1	1	2	6	4	6	4 bytes

Figure 2.4. ARP packet format.

corresponding MAC address found in the cache is used and no ARP request and reply will be sent.

Figure 2.4 shows the format of an ARP message, which is 28 bytes long. An ARP request or reply is encapsulated in an Ethernet frame, with the `Protocol Type` field set to 0x0806. An 18-byte padding is needed since the minimum length of an Ethernet frame is 64 bytes.

In Fig. 2.4, the first four fields define the types of the addresses to be resolved. `Hardware Type` specifies the type of physical address used, and `Protocol Type` specifies the type of the network protocol address. The next two fields give the length of these two types of addresses. The `Operation` field specifies whether it is an ARP request (with a value of 1), ARP reply (2), RARP request (3), or RARP reply (4). The following four fields are the MAC and IP addresses of the sender and the targeted receiver, respectively.

PPP networks do not use ARP. In this case, hosts must know the IP address at the end of the PPP link. Usually DHCP is used over a PPP link where the IP address of one end host is assigned automatically by the other end host.

Proxy ARP and gratuitous AR0

Proxy ARP and gratuitous ARP are two interesting scenarios of using ARP. Usually proxy ARP is used to hide the physical networks from each other. With proxy ARP, a router answers ARP requests targeted for a host. An example is shown in Fig. 2.5. There are several interesting observations to make in this example.

1. `Host_A` and `Host_B` are in two different subnets connected by a router. By setting their network masks appropriately, they are logically in the same subnet (directly connected) (see the example in Fig. 2.9 and Section 0.5.)

2. In the ARP table of `Host_A`, all the IP addresses in the `Host_B` subnet are mapped to the same MAC address, i.e., Router Port 0's MAC address.

3. All packets from the `Host_A` subnet to the `Host_B` subnet are sent to the router first, and forwarded by the router to the destination.

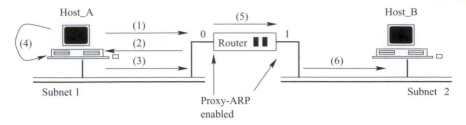

(1): Host_A broadcasts an ARP request for Host_B

(2): Router Port 0 replies for Host_B

(3): Host_A sends the frame to Router Port 0

(4): Host_A inserts a new entry in its ARP cache: {(Host_B's IP) at (Router Port 0's MAC)}

(5): Router forwards the frame to port 1

(6): Router port 1 sends the frame to Host_B

Figure 2.5. A proxy ARP example.

Gratuitous ARP occurs when a host broadcasts an ARP request resolving its own IP address. This usually happens when the interface is configured at bootstrap time. The interface uses gratuitous ARP to determine if there are other hosts using the same IP address. It also advertises the sender's IP and MAC address, and other hosts in the network will insert this mapping into their ARP table.

Manipulating the ARP table

An entry in the ARP table has three elements: (1) an IP address, (2) the MAC address associated with this IP address, and (3) flags. A normal entry expires 20 min after it is created or the last time it is referred. If an entry is manually entered, it has a *permanent* flag and will never time out. If an entry is manually entered with the **pub** key word, it has an additional *published* flag which means the host will respond to ARP requests on the IP address in this entry. If a host sends an ARP request but gets no reply, an *incomplete* entry will be inserted into the ARP table. The ARP table can be manipulated using the **arp** command. Some options of **arp** are given here.

- **arp -a**: Displays all entries in the ARP table.
- **arp -d**: Deletes an entry in the ARP table.
- **arp -s**: Inserts an entry into the ARP table.

2.3 Network interface

2.3.1 Operations of a network interface

Figure 2.6 illustrates the operations of an Ethernet card with two interfaces. Most TCP/IP implementations have a loopback interface with the IP address

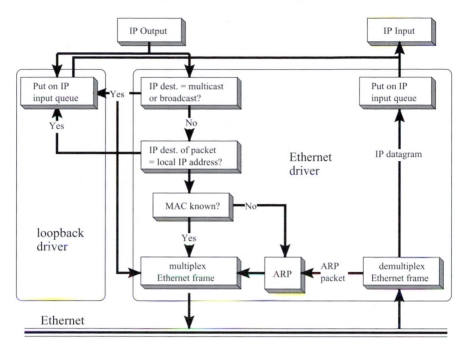

Figure 2.6. Functional diagram of an Ethernet interface card.

127.0.0.1 and domain name `localhost`. It behaves as a separate data link interface, but packets sent to it will be put in the IP input queue and will not be transmitted on the medium. The loopback interface is used for debugging.

The Ethernet interface reads and sends Ethernet frames from or to the medium. When it receives a frame, it reads the `Frame Type` field. If `Frame Type` is 0x0806, the ARP message carried in the frame is extracted and sent to the ARP module. If `Frame Type` is 0x0800, an IP datagram is extracted and sent to the IP input queue. When the IP layer has a datagram to send, the device driver first checks if it is destined for itself. In this case (i.e., multicast, broadcast, destination IP address is its own IP address, or the destination IP address is 127.0.0.1), a copy of the datagram is sent to the loopback interface. Otherwise, the device driver will search the ARP table for the destination MAC address. If there is no hit in the table lookup, an ARP request is broadcast. When the sending host gets the destination MAC address, an Ethernet frame is assembled and the device driver begins to compete for the channel, attempting to send the frame on the medium.

There is a limit on the frame size of each data link layer protocol. This limit is determined by the reliability of the physical medium and the nature of the MAC scheme used. It translates itself to a limit on the size of the IP datagram that can be encapsulated in a link layer frame, which is called the maximum transmission unit (MTU). Examples of MTUs are: 1500 bytes for Ethernet and 4352 bytes for FDDI. If an IP datagram to be sent is longer than the MTU of the interface, the IP datagram will be fragmented and carried in several data link layer frames. We will further discuss MTU and IP fragmentation in Chapter 5.

2.3.2 Configuring a network interface

The **netstat** command can be used to display the configuration information and statistics on a network interface. The same command is also used to display the host routing table. We list several **netstat** options below that will be used frequently in the following experiments.

- **netstat -a**: Shows the state of all sockets, routing table entries, and interfaces.
- **netstat -r**: Displays the routing table.
- **netstat -i**: Displays the interface information.
- **netstat -n**: Displays numbers instead of names.
- **netstat -s**: Displays per-protocol statistics.

The **ifconfig** command is used to configure a network interface. The following options are used for the reconfiguration of the IP address and network mask.

- **ifconfig -a**: Shows the states of all interfaces in the system.
- **ifconfig** *interface_name* **down**: Disables the network interface,[1] where *interface_name* is the name of the Ethernet interface.[2]
- **ifconfig** *interface_name* *new_IP_address* **up**: Assigns a new IP address to the interface and brings it up.
- **ifconfig** *interface_name* **netmask** *new_netmask*: Assigns a new network mask for the interface.

2.4 The Internet Control Message Protocol

ICMP is a protocol in the network layer that communicates error messages and other conditions that require attention, as well as routing information.

---

[1] It is recommended that the interface be disabled first, before changing its settings.
[2] Different systems may give different names for the interfaces, e.g., le0 for Sun Sparc 4 with SunOS 5.5.1 and hme0 for Sun Ultra 5 with Solaris 8. You can find the name by typing **netstat -i** or **ifconfig -a**.

Table 2.2. *Types and codes of ICMP messages*

Type	Code	Description	–
0	0	echo reply	query
3	0–15	destination unreachable	error
5	0–3	redirect	error
8	0	echo request	query
9	0	router advertisement	query
10	0	router solicitation	query
11	0–1	time exceeded	error

0	7 8	15 16	31
type(3)	code(0–15)	checksum	
unused (must be 0)			
IP header (including options), plus the first 8 bytes of the original IP datagram payload			

Figure 2.7. Format of an ICMP error message.

An ICMP message is encapsulated in an IP datagram, with the `Protocol Type` value 0x01.

There are many different ICMP messages, each of which is used for a specific task. These ICMP messages have a 4-byte common header, as shown in Fig. 2.7 and Fig. 2.8. The Type and Code fields in the common header define the function of the ICMP message. Some frequently used ICMP messages are given in Table 2.2.

Figure 2.7 displays the format of an ICMP error message. The IP header and the first 8 bytes of the payload of the original IP datagram are carried in the ICMP error message and returned to the source. The sender can analyze the returned header and data to identify the cause of the error. Note that the first 8 bytes of the original IP payload contain the source and destination port numbers in the UDP or TCP header.

Figure 2.8 gives the format of an ICMP echo request or reply message. These messages are used by the **ping** program to determine if a remote host is accessible. **ping** sends ICMP echo requests to the target IP address. The target host will respond with an ICMP echo reply for each ICMP request correctly received. The round trip time for each request/reply pair may be reported in the **ping** console. The fact that no ICMP echo reply is received means either that there is no path available to the remote host, or the remote host is not alive. When there are multiple **ping**s running on a host, each of

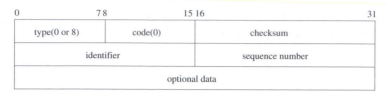

Figure 2.8. Format of an ICMP echo request or echo reply message.

them is assigned a unique identifier. The `sequence number` field is used to match reply to request.

2.5 The Sock traffic generator

Sock is a test program that can be run either as a client or as a server, using UDP or TCP. It also provides a means to set various socket options. Sock operates in one of the following four modes [5].

1. Interactive client: connects to a server, and copies the *standard input*, i.e., keys a user typed, to the server and copies everything received from the server to the *standard output*, i.e., the screen.
2. Interactive server: waits for a connection request from a Sock client, and then copies the standard input to the client and copies everything received from the client to the standard output.
3. Source client: sends packets to a specified server.
4. Sink server: receives packets from a client and discards the received data.

2.6 Network interface exercises

The following exercises use the single segment network topology shown in Fig. 1.3.

Exercise 1 | Use the **ifconfig -a** command to display information about the network interfaces on your host. Find the IP address and the net mask of your machine.

LAB REPORT   How many interfaces does the host have? List all the interfaces found, give their names, and explain their functions briefly.

LAB REPORT   What are the MTUs of the interfaces on your host?

LAB REPORT   Is network subnetted? What is the reasoning for your answer? What the experimental are the reasons for subnetting?

Exercise 2   While **tcpdump host** *your_host* is running in one command window, run **ping 127.0.0.1** from another command window.

LAB REPORT   From the **ping** output, is the 127.0.0.1 interface on? Can you see any ICMP message sent from your host in the **tcpdump** output? Why?

Exercise 3   By using **netstat -in** command, collect the statistics from all the hosts on the network. Since we use the same login name and password, we can **telnet** to other workstations and run **netstat -in** there.[3]

Save the **netstat -in** outputs.

If you don't see a significant amount of output packets in the **netstat** output, the machine was probably restarted recently. You may do this experiment later, or use the following **sock** command to generate some network traffic:

sock -u -i -n200 *remote_host* **echo**.

LAB REPORT   Calculate the average collision rate over all the hosts for the set of statistics you collected in this exercise.

2.7 ARP exercises

In the following experiment, we shall examine the host ARP table and the ARP operation, including two interesting cases: proxy ARP and gratuitous ARP. You may need to ask the lab instructor for the MAC addresses of the host and router interfaces, and record these MAC addresses in Table A.1 and Table A.2 in the appendix. You need these MAC addresses for the exercises and lab report.

Exercise 4   Use **arp -a** to see the entire ARP table. Observe that all the IP addresses displayed are on the same subnet.

If you find that all the remote hosts are in your host's ARP table, you need to delete a remote host (not your workstation) from the table, using,

arp -d *remote_host*.[4]

Save the ARP table for your lab report.

While **tcpdump -enx -w exe2.out** is running, **ping** a remote host that has no entry in your host ARP table. Then terminate the **tcpdump** program.

---

[3] After you are done with a remote host, you should exit the **telnet** session before you **telnet** to another remote host. Recursive **telnet** will generate unnecessary data in the **tcpdump** output and cause confusion.

[4] If you deleted your workstation's IP address from the ARP table by mistake, you must add the entry back in the table. See the **arp** manual page to add. Note that, in order for your workstation to reply to the ARP requests, the ARP entry of your workstation must have the **P** flag in the ARP table.

Next, run **ethereal -r exe2.out&** to load the **tcpdump** trace file.

Observe the first few lines of the packet trace to see how ARP is used to resolve an IP address.

Run **arp -a** to see a new line added in your host's ARP table. Save the new ARP table for your lab report.

Mark the ARP request packet and the ARP reply packet in the **ethereal** window. Then go to menu File/Print ... to print the marked packets for your lab report (See Exercise 6 of Chapter 1).

LAB REPORT   From the saved **tcpdump** output, explain how ARP operates. Draw the format of a captured, ARP request and reply including each field and the value.

Your report should include the answers for the following questions.
- What is the target IP address in the ARP request?
- At the MAC layer, what is the destination Ethernet address of the frame carrying the ARP request?
- What is the frame type field in the Ethernet frame?
- Who sends the ARP reply?

Exercise 5   While **tcpdump host** *your_host* is running to capture traffic from your machine, execute **telnet 128.238.66.200**. Note there is no host with this IP address in the current configuration of the lab network.

Save the **tcpdump** output of the first few packets for the lab report.

After getting the necessary output, terminate the **telnet** session.

LAB REPORT   From the saved **tcpdump** output, describe how the ARP timeout and retransmission were performed. How many attemps were made to resolve a non-existing IP address?

Exercise 6   The network topology for this proxy ARP exercise is shown in Fig. 2.9. We will divide the group into two subnets interconnected by a router. The IP addresses and network masks for the hosts are also given in Fig. 2.9. Change the IP address and network mask of your host accordingly (see Section 2.3.2). The IP addresses and network masks of the Router4 interfaces are the same as their default settings. Note that the network mask of the hosts in the 128.238.65.0 network is 255.255.0.0.

Next we will enable the proxy ARP function on the ethernet1 interface of Router4.
1. **telnet** to Router4 from shakti: **telnet 128.238.64.4**. The login password is e1537.[5]

---

[5] Check with your lab instructor for the password of the router your are using, which may be different from e1537.

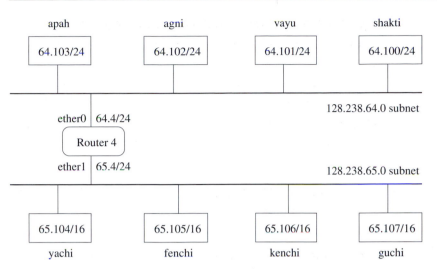

Figure 2.9. Network configuration for the proxy ARP experiment.

2. Log in to the router, type **enable** to enter the *Privileged EXEC* mode.[6] The password is again `e1537`.
3. Enter the *Global Configuration* mode by typing **config term**.
4. Then type the following lines:
> **interface ethernet 1**[7]
> **ip proxy-arp**
> `Ctrl-Z`
5. Type **exit** to terminate the **telnet** session.

Now `Router4`'s `ethernet1` interface can perform proxy ARP for the hosts in the 128.238.64.0 subnet.

Run **tcpdump -enx** on all the hosts.

Then let the hosts in the 128.238.65.0 subnet send UDP datagrams to the hosts in the 128.238.64.0 subnet. For example, on `guchi` type:

> **sock -i -u -n1 -w1000** *Host_in_64.0_subnet* **echo**.

When you are done with all the hosts in the 128.238.64.0 subnet, save the tcpdump output for the lab report.

Run **arp -a** to display the new ARP table in your host. Save the ARP table for your lab report.

After the lab instructor restores the network into a single subnet (see Fig. 1.3), change the IP address and network mask of your host's interface back to their default values as in Fig. 1.3.

---

[6] We will discuss bridge and router configuration in Chapter 3.
[7] The name of the router interfaces may be different for various routers. You can find the names by typing **write term** in the *Privilege EXEC* mode.

Exchange your data saved in this exercise with a student working in the other subnet.

LAB REPORT   Explain the operation of proxy ARP.

Why can a host in the 128.238.65.0 subnet reach a host in the 128.238.64.0 subnet, even though they have different subnet IDs?

What are the MAC addresses corresponding to hosts in the 128.238.64.0 subnet, in the ARP table of a host in the 128.238.65.0 subnet?

Give one advantage and one disadvantage of using proxy ARP.

Exercise 7   This exercise will be performed by all the students together. While **tcpdump -ex -w exe7.out** is running on all the hosts, reboot host guchi.

After guchi is started, terminate **tcpdump** and run **ethereal -r exe7.out &** to load the **tcpdump** trace. Print the the gratuitous ARP request for your lab report.

LAB REPORT   What is the purpose of gratuitous ARP?

LAB REPORT   List the sender IP address, target IP address, sender MAC address, and target MAC address of the gratuitous ARP you saved.

2.8 Exercise with ICMP and Ping

Exercise 8   Use **ping -sv** *remote_host* to test whether the remote host is reachable, while running: **tcpdump -enx host** *your_host* **and** *remote_host*.

Save the **tcpdump** and **ping** output for the future study on **ping**.

LAB REPORT   What ICMP messages are used by **ping**?

Exercise 9   While running **tcpdump -x -s 70 host** *your_host* **and** *remote_host*,

execute the following **sock** command to send a UDP datagram to the remote host: **sock -i -u -n1 -w1000** *remote_host* **88888**.

Save the **tcpdump** output for the lab report.

LAB REPORT   Study the saved ICMP port unreachable error message (see Fig. 2.7). Why are the first 8 bytes of the original IP datagram payload included in the ICMP message?

Exercise 10   While **tcpdump** is running to capture the ICMP messages, **ping** a host with IP address 128.238.60.100. Save the **ping** output.

Table 2.3. *Host IP addresses and network masks for exercise 11*

Group	Name	IP address	Subnet mask
Group A	shakti	128.238.66.100	255.255.255.0
	vayu	128.238.66.100	255.255.255.0
	agni	128.238.66.102	255.255.255.0
	apah	128.238.66.103	255.255.255.0
Group B	yachi	128.238.66.104	255.255.255.0
	fenchi	128.238.66.104	255.255.255.0
	kenchi	128.238.66.106	255.255.255.0
	guchi	128.238.66.107	255.255.255.0

LAB REPORT   Can you see any traffic sent on the network? Why? Explain what happened from the **ping** output.

LAB REPORT   List the different ICMP messages you captured in Exercises 8, 9, and 10 (if any). Give the values of the type and code fields.

2.9 Exercises with IP address and subnet mask

In this section, we will observe what happens when the same IP address is assigned to two different hosts. We will also set an incorrect subnet mask for hosts and see what are the consequences. For the next two exercises, we split the current single segment network into two segments, Group A and Group B as shown in Table 2.3, so that they will not interfere with each other.

Exercise 11   Change the IP address of your workstation as shown in Table 2.3.

Delete the entries for all hosts other than your own workstation from your workstation's ARP table.

Run **tcpdump -enx** on all the hosts. Then, do the following three experiments.
1. Execute **telnet** from one of two hosts with the duplicate IP address to a host with unique IP address (e.g. shakti –> agni in Group A and yachi –> kenchi in Group B).
Now, from the other host with the duplicate IP address, execute **telnet** command to the same host (vayu –> agni or fenchi –> kenchi).
Observe what happens and save the **tcpdump** output and the ARP tables in all the hosts in your group.

Table 2.4. *Host IP addresses and network masks for exercise 12*

Group	Name	IP address	Subnet mask
Group A	shakti	128.238.66.100	255.255.255.240
	vayu	128.238.66.101	255.255.255.0
	agni	128.238.66.102	255.255.255.0
	apah	128.238.66.120	255.255.255.240
Group B	yachi	128.238.66.104	255.255.255.240
	fenchi	128.238.66.105	255.255.255.0
	kenchi	128.238.66.106	255.255.255.0
	guchi	128.238.66.121	255.255.255.240

2. Execute **telnet 128.238.66.100** (or **128.238.66.104**) from agni (or kenchi). Which host provides the telnet connection? Why?
3. Execute **telnet 128.238.66.100** (or **128.238.66.104**) from apah (or guchi). Which host is connected to apah (or guchi)? Why?

LAB REPORT   Explain what happened in the first case and why. Answer the questions for the second and third cases.

Exercise 12   Change the host IP addresses and the subnet masks as shown in Table 2.4. Since we still have two separate segments, Groups A and B can do the exercise independently. Note that two hosts in each group (shakti and apah in Group A, or yachi and guchi in Group B) are assigned an incorrect subnet mask.

Capture the packets with **tcpdump -e** for the following cases.
1. When shakti (yachi) **ping**s one of the hosts that have the correct subnet mask.
2. When apah (guchi) **ping**s one of the hosts that have the correct subnet mask. Now, copy the output displayed from the **ping** window in apah (guchi). Share the saved output message with other students.
3. When a host with the correct subnet mask **ping**s shakti (yachi).
4. When a host with the correct subnet mask **ping**s apah (guchi).
To avoid confusion, only one machine in each group should generate traffic in each case. Clearly, this exercise has to be performed as a team.

LAB REPORT   Explain what happened in each case according to the **tcpdump** outputs saved. Explain why apah (or guchi in Group B) could not be reached from other hosts, whereas shakti (or yachi in Group B), which has the same incorrect subnet mask, could communicate with the other hosts.

3 Bridges, LANs and the Cisco IOS

Algorhyme

I think that I shall never see
A graph more lovely than a tree.
A tree whose crucial property
Is loop-free connectivity.
A tree that must be sure to span
So packets can reach every LAN.
First, the root must be selected.
By ID, it is elected.
Least-cost paths from root are traced.
In the tree, these paths are placed.
A mesh is made by folks like me,
Then bridges find a spanning tree.

Radia Perlman

3.1 Objectives

- The Cisco Internet Operating System software.
- Configuring a Cisco router.
- Transparent bridge configuration and operation.
- The spanning tree algorithm.

3.2 Ethernet bridges

3.2.1 Use of bridges

Bridges are link layer devices. As illustrated in Fig. 0.3, when two network segments with different link and physical layer protocols are connected, the bridge performs a two-way translation of the protocols. The data section of a transit frame is extracted and re-encapsulated in the frame format used by the next-hop network segment.

However, there are several reasons to use bridges to connect networks even with identical protocols, rather than using a large network without bridges. First, network segments could be far away from each other but still work within the same logical network. In this case, two remote network segments can be linked by two bridges via a point-to-point wide area link. Second, there is a limit on the maximum length of the shared medium in a local network. This further limits the physical size of a single segment network. Using bridges can effectively extend the size of a local area network. Third, since the throughput of a local network decreases as the collision rate increases (see exercises in Chapter 2), small LAN segments always perform better than large segments with more devices. Last, breaking a local network into several segments connected by bridges has security advantages, since communications within a segment cannot be overheard from outside that segment.

We will focus on the IEEE 802.1d bridge (which is widely used) in discussing bridge operations in the following.

3.2.2 Bridge operation

A bridge has several ports, each connected to a network segment. As an internetworking device, a bridge's basic function is to forward frames from one LAN segment to another. When a *transparent bridge* is used, a frame is simply copied to the destination network, with no modification in the header and data section, and the end devices are not aware of the presence of bridges.

To perform the forwarding function, MAC addresses of the hosts stored in a *filtering database* in the bridge is used. The filtering database consists of a number of entries, each with three elements: (1) the destination MAC address, (2) the bridge port where frames for this destination MAC address should be forwarded to, and (3) the age of this entry. The filtering database could be set manually and be static. However, in an IEEE 802.1d bridge, the filtering database is maintained automatically by an address learning process, as illustrated in Fig. 3.1. When the bridge receives a frame from one of its ports, it infers that the source of this frame can be reached from the incoming port. Then, the source MAC address field of the frame and the port needed to reach it are updated in the bridge's filtering database. The default age of a new entry is 300 seconds, after which the entry is deleted.

A bridge makes forwarding decisions by filtering database lookups. If there is an entry corresponding to the destination MAC address of the frame

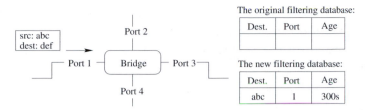

Figure 3.1. The bridge learns the source address from an incoming frame.

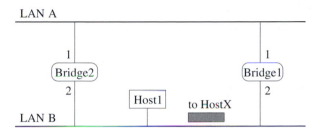

Figure 3.2. When there is a loop in the bridged network, the address learning scheme will not work.

found, the bridge forwards this frame to the network segment indicated by the entry. Otherwise, flooding is used where the received frame is copied to all the active ports except the incoming port.

3.2.3 Spanning tree algorithm

The address learning scheme described in the last subsection works fine if there is no loop in the network. When there are more than one path between a source and a destination, as shown in the example in Fig. 3.2, the address learning and forwarding scheme may cause serious problems.

In Fig. 3.2, suppose Host1 sends a frame to a HostX (which is not shown in the figure), for which there is no entry in Bridge1 and Bridge2's filtering database. Both bridges receive the frame on LAN B, and learn that Host1 is on LAN B. So each of the bridges correctly update (or add) the entry for Host1 in its filtering database. Also, both bridges forward the frame to LAN A using flooding since there is no entry for HostX in their filtering database. Then, each bridge will receive the same frame forwarded by the other bridge, and will incorrectly change the filtering database entry to indicate that Host1 is on LAN A. Next, each bridge will forward the frame on LAN B again. This process, which repeats indefinitely, is known as a

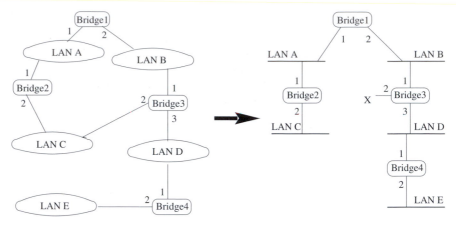

Figure 3.3. An example bridged network with a loop and the corresponding tree with the loop removed.

broadcast storm. A severe broadcast storm can block other network traffic, resulting in a network meltdown.

The solution to this problem is to remove loops in bridged networks. A bridged network can be viewed as a graph, where the bridges are nodes and the network segments are edges. A tree is a graph with no loops. If we can build a tree from this graph by disabling some of the bridge ports, loops will be removed. This is shown in Fig. 3.3, where the loop in the network is removed by disabling port 2 of Bridge3. It can be seen that from any host in the tree network to any other host, there is only a single path. The problem discussed in the previous example is solved.

The spanning tree algorithm defined in the IEEE 802.1d standard is used in bridged networks to build trees dynamically. It works as follows.

1. Each bridge is assigned a unique identifier, and each port of a bridge is assigned an identifier unique to that bridge. Typically, the identifier of a bridge is a priority concatenated with one of the bridge ports' MAC address, and the identifier of a port is a priority concatenated with a port index local to the bridge. Each bridge port has a corresponding path cost, which indicates the cost to transfer a frame to an attached network segment through that port.

2. Select the *root bridge*, which is the one with the lowest-value bridge identifier. The ID of the root is called the *root ID*.

3. Each bridge selects its *root port*. The root port of a bridge is the port from which the root bridge can be reached with the least aggregate path cost (called the *root path cost*).

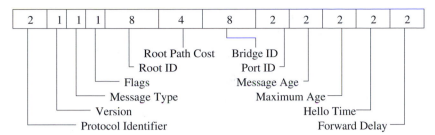

Figure 3.4. BPDU message format. The numbers indicate the field length in byte.

4. Determine the *designated bridges* and the *designated ports*. Each network segment is associated with a designated bridge, which provides the shortest path to the root bridge and is the only bridge allowed to forward frames to and from the root. The port connecting a designated bridge to the network segment is a designated port. If more than one bridge provides the same root path cost, the bridge with the lowest-valued bridge identifier is selected as the designated bridge.

5. Only the root ports and designated ports of the bridges are allowed to forward frames. All other bridge ports are blocked.

6. The above steps are repeated whenever the network topology changes.

In the tree shown in Fig. 3.3, `Bridge1` is the root bridge since it has the smallest bridge identifier. `Bridge4`'s root port is port 1, with a root path cost of 2 hops. `LAN D`'s designated bridge is `Bridge3`, and its designated port is `Bridge3`'s port 3.

To implement the spanning tree algorithm in a distributed manner, bridges exchange configuration information using a message called *bridge protocol data units* (BPDUs). The format of a BPDU message is given in Fig. 3.4 with the definition of the fields given below.

- `Protocol Identifier`, `Version`, and `Message Type`: These three fields are always set to 0.
- `Flags`: The least significant bit, called the Topology Change (TC) bit, is set to signal a topology change. The most significant bit is to acknowledge receipt of a BPDU with the TC bit set. The remaining six bits are not used.
- `Root ID`: Identifies the root bridge by listing its 2-byte priority followed by a 6-byte Ethernet address. The priority value can be set in the *Global Configuration* mode. The default priority is 0x8000.
- `Root Path Cost`: The path cost to the root bridge.
- `Bridge ID`: The identifier of the bridge sending the message.

Figure 3.5. Cisco IOS enables network applications on the network platforms.

- `Port ID`: Each bridge port has a unique 2-byte identifier. The first byte is the priority, which is configurable, while the second byte is a number assigned to the port.
- `Message Age:`[1] Specifies the amount of time since the root originally sent the BPDU on which the current configuration message is based.
- `Maximum Age:`[1] Indicates when the spanning tree topology is recalculated if a bridge does not hear BPDUs from the root bridge. The default value is 15 seconds.
- `Hello Time:`[1] Provides the time period between two BPDUs from the root bridge. The default value is 1 second.
- `Forward Delay:`[1] provides the amount of time that bridges should wait before switching a port from the blocking state to forwarding state. If a bridge port switches state too soon, not all network links may be ready to change their state, and loops can occur. The default value is 30 seconds.

3.3 Configuring a bridge or router

3.3.1 The Cisco internet operating system

For a bridge or router to work properly, we need higher layer functions for configuration and management tasks. The Cisco IOS is the most widely deployed network system software, delivering network services such as operations, administration, and maintenance of the network platforms and Internet applications. Cisco IOS supports a broad range of platforms, as well as many networking protocol families, such as TCP/IP, AppleTalk, DECnet, Systems Network Architecture (SNA), among others. As shown in Fig. 3.5, Cisco network platforms and the Cisco IOS software running on them are a unified system.

---

[1] in 1/256ths of a second.

In our lab, the Cisco IOS software is running in the four routers. In the following exercises, we will learn how to use the Cisco IOS to configure a bridge or a router. More specifically, we will configure the interfaces, enable or disable different functions, choose what protocol (e.g., the spanning tree algorithm, RIP or OSPF) to use, and display the state of the bridge/router.

3.3.2 Cisco IOS configuration modes

The Cisco IOS provides different ways to configure and maintain a Cisco device. The Cisco IOS command-line interface (CLI) is the primary user interface which allows you to directly and simply execute Cisco IOS commands, whether using a router console or terminal, or using remote access methods. The Cisco IOS software also includes a web browser user interface (UI) from which you can issue Cisco IOS commands. The Cisco IOS web browser UI can be accessed from the router home page.

There are six different configuration modes in CLI, each providing a set of configuration commands. A set of special commands is used to navigate through these modes. The six configuration modes are listed below.

1. *User EXEC* mode. This is the mode you are in after you login to the router. The EXEC commands available in this mode are a subset of those in the privileged EXEC mode. Most commands in this mode are used to determine the router status, but do not change the configuration of the router.

2. *Privileged EXEC* mode. This is the second level of access for the EXEC mode. All the EXEC commands are available in this mode. This mode provides access to the configuration mode by means of the **configure** command, and includes advanced testing commands, such as **debug**.

3. *Global configuration* mode. This mode provides commands to configure the system globally. Global attributes can be configured in this mode.

4. *Interface Configuration* mode. This mode provides commands to configure an interface. Attributes that relate to an interface can be set in this mode.

5. *Subinterface Configuration* mode. This is a submode of the interface configuration mode. In this mode you can configure multiple virtual interfaces (called subinterfaces) on a single physical interface.

6. *ROM Monitor* mode. This mode is used to manually locate a valid system software image from which the device bootstraps.

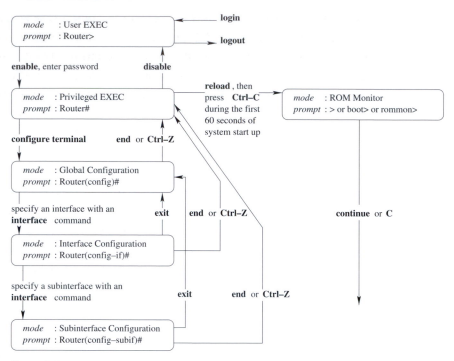

Figure 3.6. Navigating through the Cisco IOS configuration modes.

In the following experiments, we will use commands in the first four configuration modes. Figure 3.6 further illustrates configuration modes, their prompts, and commands used to navigate through them.

To get help, typing **?** displays all the commands available in the mode you are in. Typing **?** after a partial command string lists commands in the current mode that begin with that string. Typing **?** after a full command lists the available syntax for that command.

3.3.3 Bridge/router configuration procedure

As shown in Fig. 3.6, you need to log into the router to perform configurations. If the network is not available, you can configure a router locally. Each router has a serial console port. You can directly connect a console terminal (an ASCII terminal or a terminal emulator) to the console port of the router (see Section A.3.3).

When the router and a host are in the same LAN, you can **telnet** to the router for configurations. In order to access the router, you should make sure that your host is in the same subnet as the router interface. If the workstation

IP address doesn't match the subnet of a router, you first need to set your workstation IP address to match the subnet of the router that you want to configure. Then you can **telnet** to the router interface and change the IP address of each interface as required. If you change the router interface to another subnet, the **telnet** session will be frozen since now the router interface is in a different subnet from your host. In this case, kill the window and change the IP address of your workstation again to match that of the router interface. Again, **telnet** to the new IP address of the router interface (which you just set), and then do the remaining configuration.[2]

The following is an example of router configuration from a remote host.

1. Connect the host and a router interface using a hub. Also change the host's IP and/or netmask to match the subnet of the router interface if necessary.

2. **Telnet** to the router interface, enter the virtual terminal password, which is e1537. Now you are in the *User EXEC* mode with prompt:

       Router>

3. You may enter the *Privileged EXEC* mode by typing:

       Router> **enable**

   After entering the enable password (e1537), you will see the *Privileged EXEC* prompt:

       Router#

   You may type **write terminal** to display the current configuration in the *Privileged EXEC* mode.

4. To begin a new configuration, use the following command in the *Privileged EXEC* prompt:

       Router# **configure terminal**

   Then the router displays an explanation of the editing functions. Now you are in the *Global Configuration* mode.

   In this mode, you can make various configurations, e.g., assigning IP addresses to the interfaces, specifying which protocols to run, etc. Since it is impossible to describe all the configuration commands here, we will introduce configuration commands whenever they are needed.

5. To end the configuration mode, type **Ctrl-z**. Then, you will get back to the *Privileged EXEC* mode. You can examine the new configuration by typing **write terminal** again.

---

[2] By defining a virtual interface called *loopback*, we can change the IP address of any router interface without an interrupt of the **telnet** session. However, for the sake of experiments, we do not introduce the loopback feature in this lab.

6. Enter the **disable** command to return to the *User EXEC* mode. Type **exit** to end the connection to the router.

Here, we intentionally omit the command for saving the configuration changes to NVRAM (Nonvolatile RAM). For more information about the configuration commands, see [6] and [7].

3.3.4 Configuring a transparent Bridge

This section discusses how to set the IP address of a bridge or router interface and how to initiate the transparent bridge function. The following procedure sets the IP address of an interface in a bridge or router.

1. Get to the *Global Configuration* mode (see the previous section, or Fig. 3.6).
2. To specify an interface and start the interface configuration, enter:
   Router(config)# **interface** *interface_type interface_number*.
   For example, type **interface ethernet 0** to configure the ethernet 0 interface. Note that since you are in the *Global Configuration* mode, you have access to all the bridge/router interfaces (e.g., ethernet 1), and can configure them here.
3. From the *Interface Configuration* mode, use the command:
   Router(config-if)#**ip address** *new_IP_address net_mask*
   to set the IP address and subnet mask of the interface.
4. Type **Ctrl-z** to return to the *Privileged EXEC* mode. You can type **write terminal** to verify the changes just made.
5. Type **disable** to return to the *User EXEC* mode. From both EXEC modes, you can type **exit** to terminate the telnet session.

To configure a transparent bridge with the spanning tree algorithm, the following steps are needed in the *Global configuration* mode.

1. To bridge (as opposed to route) IP datagrams, disable IP routing:
   Router(config)# **no ip routing**
2. Enable each interface:
   Router(config)# **interface ethernet 0** or **1**
3. Assign the network interfaces to a spanning-tree group:
   Router(config-if)# **bridge-group** *group*
   The argument *group* is a number between one and nine that you choose to refer to a particular set of bridged interfaces on this bridge. Frames are bridged only among interfaces in the same group.

Table 3.1. *A basic transparent bridge configuration example*

no ip routing
interface ethernet 0
ip address 128.238.61.1 255.255.255.0
bridge-group 1
interface ethernet 1
ip address 128.238.61.2 255.255.255.0
bridge-group 1
bridge 1 protocol ieee

4. Define the spanning-tree protocol:

    Router(config-if)# **bridge** *group* **protocol** *protocol*

protocol specifies which spanning tree protocol to use. It can be either **ieee** for the ieee spanning-tree protocol, or **dec** for the DEC spanning-tree protocol.

5. Set the other spanning-tree parameters, e.g., the bridge priority and path costs, if necessary.

Table 3.1 is an example of a basic transparent bridge configuration used in Exercise 3 of this chapter.

Initially, routers were configured as shown in Appendix A. Unless you save the configuration changes to NVRAM permanently, the router will always reboot with that configuration.

3.4 Exercises on Cisco IOS

The students in the lab should divide themselves into four groups for the first four exercises. Each group uses two workstations, a bridge, and two hubs, which are required to be connected as shown in Fig. 3.7, Table 3.2 and Table 3.3.

Exercise 1   In this exercise we build the connection to the router.

Identify the cable from your workstation and the cable from your router interface (see Fig. 3.7, Table 3.2 and Table 3.3). Plug these two cables into your hub. In this case, you have built a LAN segment with a *star* topology. Your partner should build a star LAN segment on the other side of the router.

Table 3.2. *Router IP addresses for Fig. 3.7*

	eth0	eth1
router1	128.238.61.1/24	128.238.61.2/24
router2	128.238.62.2/24	128.238.62.3/24
router3	128.238.63.3/24	128.238.63.4/24
router4	128.238.64.4/24	128.238.64.5/24

Table 3.3. *Host IP addresses for Fig. 3.7*

HOST_A			HOST_B		
Name	IP address	label	Name	IP address	label
shakti	128.238.61.101/24	1	vayu	128.238.61.102/24	2
agni	128.238.62.102/24	3	apah	128.238.62.103/24	4
yachi	128.238.63.103/24	5	fenchi	128.238.63.104/24	6
kenchi	128.238.64.104/24	7	guchi	128.238.64.105/24	8

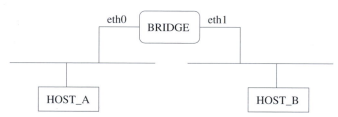

Figure 3.7. Using a transparent bridge.

After running **tcpdump -enx** on both workstations, turn on the router. Capture the gratuitous ARP sent by the router.

Change the IP address of your workstation to be in the same subnet as the router. You can choose any valid host id for your host.

ping the router interface to test the connection to the router.

LAB REPORT   Submit the gratuitous ARP sent by the router. What is the default IP address of the router interface?

Exercise 2   **Telnet** to your router. When prompted for a login password, type e1537. You should now be in the *User EXEC* mode.

Type **help** to learn how to use the online help.

Study Fig. 3.6. Navigate through the *User EXEC*, *Privileged EXEC*, *Global Configuration*, and *Interface Configuration* modes. In each mode, type **?** to display a list of available commands and study these commands.

Type **show version** in the *User EXEC* mode to display the Cisco IOS banner. Identify which Cisco IOS Release is running in the router. Save the Cisco IOS banner for your lab report.

LAB REPORT   Submit the Cisco IOS banner you saved. Identify the release of the Cisco IOS software in the router.

3.5 A simple bridge experiment

Figure 3.7 shows a simple case of the use of bridges, which consists of two network segments connected by a bridge. With this simple topology, we can easily capture initial BPDUs before each bridge is engaged in the spanning tree calculation.

Configure transparent bridging as in Fig. 3.7, Table 3.2 and Table 3.3. Note that the default configuration of the hosts and the bridges are different from those in the tables. You need to change the IP addresses of the bridge interfaces,[3] as well as set the bridge group and enable the spanning tree algorithm (see the previouse section on bridge configuration). Do the following experiments.

Exercise 3 | Configure the IP addresses of your workstation and the bridge interfaces as shown in Fig. 3.7, Table 3.2 and Table 3.3. To avoid confusion, each bridge should be configured by only one person.

Run **tcpdump -en ip proto 1** on your machine, and your partner's machine.

Send **ping** messages to your partner's machine: **ping -sv** *remote_machine*.

After receiving the tenth echo reply, quit the **ping** process, and save the **tcpdump** outputs from both machines.

During this exercise, don't run **ping** programs at the same time. For clean results, do your experiments in turn.

LAB REPORT   What are the IP and MAC addresses of a packet that went from your machine to the bridge? What are the IP and MAC addresses of a packet that went from the router to your partner's machine?

---

[3] As soon as you change the IP address of the bridge interface your host is connected to, the **telnet** connection will be lost. You need to again change the IP address of your workstation to be in the same subnet as the bridge interface. See Section 3.3.3.

Answer the same questions, but for the echo reply that was returned from your partner's machine.

LAB REPORT   Using the **tcpdump** outputs from both machines, calculate the average delay that a packet experienced in the bridge. Note that the system times of the two machines might be different. Show all the steps and submit the **tcpdump** outputs with your report.

Exercise 4   Run **tcpdump -e -c 5 ether multicast** on your workstation to capture 5 BPDUs messages generated by the bridge. Save the BPDUs for the lab report.

You need to collect all the different BPDUs from other students in your lab. At this time, however, just save your BPDU in the guest home directory (which is /home/guest/) as "*name_of_your_host*.ex4" since there is no network connections to hosts in the other groups. In our next exercise, after we put all the workstations in one network as shown in Fig. 3.8, you can collect BPDUs from other workstations using **ftp**.

You should collect eight different BPDUs in this exercise. These BPDUs will be helpful when studying the spanning tree algorithm later in this chapter.

LAB REPORT   How frequently (in seconds) does a bridge sends its BPDUs?

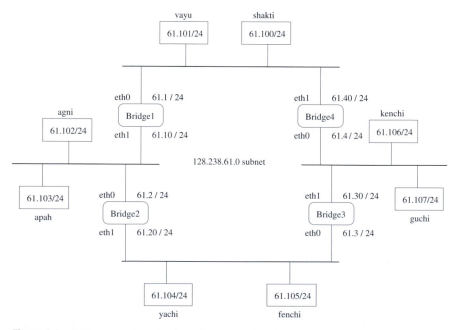

Figure 3.8. Bridge experiment network.

LAB REPORT   Submit the eight different BPDUs you saved. Identify the values of root
ID, root path cost, bridge ID, and port ID for each BPDU[4].

3.6 Spanning tree exercises

In this section, we will use Fig. 3.8 as our network topology. You need
to change the IP addresses of the bridge interfaces, as well as that of your
workstation. Refer to Section 3.3.4 on how to configure a transparent bridge.
Also see Section 3.3.3 on how to handle a frozen telnet session after you
change the bridge IP address.

Upon being started, a transparent bridge learns the network topology by
analyzing source addresses of incoming frames from all attached networks.
The next exercise shows the process by which a transparent bridge builds
its filtering database.

Exercise 5   After configuring the network in Fig. 3.8, login to the bridge.

Get to the *Privileged EXEC* mode. Type **show bridge** to see the entries in the bridge
forwarding database.

Whenever you **ping** or **telnet** from your workstation to a host that is not in the table,
observe how the filtering database in the bridge is expanded.

You may use the **clear bridge** *group* command to remove any learned entries from
the filtering database, if you see a full filtering database or if you want to repeat the
above exercise.

LAB REPORT   From the output of **show bridge**, identify which bridge ports are blocked,
and which ports are in the forwarding state for each bridge.

Exercise 6   Using **tcpdump -ex ether multicast**, capture the BPDU packet flowing on your
network segment.

Telnet to the hosts in the other three LAN segments and execute the above **tcpdump**
command in the **telnet** window to collect BPDUs sent there.

Login to each bridge to collect the **show bridge** outputs.

LAB REPORT   Submit the four different BPDUs you saved. Identify the values of root
ID, root path cost, bridge ID, and port ID for each BPDU.

LAB REPORT   Based upon the initial BPDUs saved in Exercise 4, draw the spanning
tree seen by the BPDUs. Identify the root ports and the root path cost (in hop

---

[4] You may ask the lab instructor for the physical addresses of network interfaces, and record them in
Table A.1 and Table A.2. You need the MAC addresses to help analyze the BPDUs.

counts) for each bridge. Identify the designated bridge and the designated port for each LAN segment. Identify the state of each bridge port (blocking or forwarding).

Don't just assume that `Bridge1` has the highest priority for the root bridge. Draw the spanning tree based upon your data (eight initial BPDUs).

Write the final BPDUs you collected using the three-tuple format: {*root ID, root path cost, bridge ID*}.

Once you have the spanning tree, justify it using the four final BPDUs collected in this exercise and/or the output of the **show bridge** command.

| Exercise 7 | This exercise is performed by all the students together. First, send **ping** messages from `apah` to `yachi`, while **tcpdump** is running. Let the two programs run during this exercise.

Then, disconnect the cable from the `ethernet0` port of `Bridge2` from the hub, and type the **time** command on `apah` or `yachi` to get the current time.

Observe the **ping** and **tcpdump** windows. When the connection is reestablished, type the **time** command again. How long does it take the spanning tree algorithm to react to the change in the topology?

Once you can successfully reach other hosts, get to the bridges to run **show bridge** to collect the port states. Also collect BPDUs from all the LAN segments as you did in the previous exercise.

After every student has collected the required data, connect the cable to the original position. Again, measure the time it takes for the bridges to adapt to the new change.

LAB REPORT   Draw the new tree formed after the cable was disconnected, based on the BPDUs you collected in this exercise. Specify the state of each bridge port.

3.7 Exercise on the Cisco IOS web browser UI

| Exercise 8 | You can also configure a router using the web browser UI. To enable the web server, login to the router and execute **ip http server** in the *Global Configuration* mode.

Next, start a web browser (e.g., `Mozilla` in Linux, or `Hotjava` in Solaris) in your host, and enter the IP address of the router interface. When prompted, enter `el537` for password, and leave the `User Name` field blank. Then you can browse the router configuration web pages and configure the router there.

Static and dynamic routing

We hoped that we could find a way to permit an arbitrary collection of packet-switched networks to be interconnected in a transparent fashion, so that host computers could communicate end-to-end without having to do any translations in between.

Vinton G. Cerf

4.1 Objectives

- Comparison of router and bridge.
- IP forwarding.
- Use of ICMP messages in routing.
- The Routing Information Protocol (RIP).
- The Open Shortest Path First (OSPF) protocol.
- Static routing by manually building the routing tables in the routers and hosts.
- Use of Traceroute to find an end-to-end route.

4.2 Static and dynamic routing

Routing is the act of transferring packets from a source to a destination using network layer protocol information. It involves two activities, determining optimal routing paths and transporting packets through an internetwork. The key to these two activities is the routing table maintained in each host and router. The routing table records optimal routes and is consulted when a forwarding decision is to be made for an arriving packet. The routing table can be manually set, updated by an ICMP message received, or by routing daemons implementing dynamic routing protocols.

4.2.1 Next-hop routing

When a host needs to send a packet to a destination, it uses and the IP address of the destination host to find out the network and ID (or the extended prefix) of the destination. If the host and the destination have the same extended prefix, the host and destination are on the same network. Then the host sends the datagram directly to the destination. This is known as *direct delivery*, as shown in Fig. 0.3. If the extended prefix of the host is different from that of the destination, the host and the destination are in different networks. The host must use *indirect delivery* and send the datagram to a router, as shown in Fig. 0.5. The router is then responsible for delivering the datagram to its destination.

The routing table is consulted for each *indirect delivery* in order to determine the next hop router. Only one hop on the path from the router to a destination network is listed in the routing table, instead of the whole path. Each entry in the routing table points to a router to which it connects directly. More specifically, each entry in the routing table contains the following.

- A destination IP address: either a complete host address or a network address. A host address has a nonzero host ID, while a network address has a host ID of 0.
- The IP address of the next-hop router, or of a directly connected network.
- Flags. There are five flags which can be used for a given route.
  - U. The route is up.
  - G. The route is to a router (gateway).
  - H. The route is to a host.
  - D. The route was created by a *redirect* (see Section 4.2.3).
  - M. The route was modified by a redirect.
- The host's network interface that the datagram should be delivered to, e.g., the host's Ethernet interface eth0.

When a router receives a datagram, it extracts the destination IP address and computes the network prefix. Then the forwarding decision is made according to the result of routing table lookup, as follows.

1. If the network prefix matches any directly connected network address, the datagram is delivered directly to the destination over that network.
2. Else if the table contains a host-specific route for that address, the datagram is sent to the next hop router specified in that table.
3. Else if the table contains a network-specific route for the destination host's subnet, the router forwards the datagram to the router of that network.

Table 4.1. *An example host routing table (Red Hat Linux 9)*

Destination	Gateway	Genmask	Flags	MSS	Window	irtt	Iface
128.238.4.0	0.0.0.0	255.255.255.0	U	40	0	0	eth0
127.0.0.0	0.0.0.0	255.0.0.0	U	40	0	0	lo
0.0.0.0	128.238.4.1	0.0.0.0	UG	40	0	0	eth0

4. Else if there is a default router entry in the routing table, the datagram is sent to the default router.
5. If not even a default router is found, a routing error is generated and the datagram is dropped.

As shown above, host entries have priority over network entries, which have priority over default entries. This sequence of lookups is called the *longest-prefix-matching* rule and is commonly used in routing table lookup. Table 4.1 gives an example routing table from a Linux machine, where the first entry is for the host's own subnet, the second entry is for the loopback interface, and the third entry is the default route with a default router 128.238.42.1 and a G flag. Both the first and the third entry use the local Ethernet interface eth0, while the loopback entry uses the loopback interface lo. When the host has a packet to sent to a destination in the 128.238.4.0 subnet, both the first and the third entry match the destination address. However, due to the longest-prefix-matching rule, the first route is used. If the host has a packet to a destination of 128.238.66.100, the default route will be used.

Considering the fact that table lookup is performed for each IP datagram in each router along its path, and the tremendous volume of IP datagrams in today's Internet, a smaller routing table, which shortens the lookup time, is always preferred. Most routing tables do not contain host-specific entries but only network-specific entries, which keeps the table small. For hosts that can access only one router, using a default route for all the networks that are not directly connected is more efficient.

4.2.2 Static routing versus dynamic routing

Static routing is useful in the following three situations: the network is small, there is a single connection point to other networks and there are no redundant routes. Otherwise, dynamic routing is preferred.

With static routing, the routing table entries are created by default when the interface is configured during bootstrap (e.g., using the Dynamic

Host Configuration Protocol (DHCP)), added by the **route** command (from the system bootstrap file) or created by an ICMP *redirect* or *router discovery*. The latter two will be discussed in Section 4.2.3. In dynamic routing, a router communicates with other routers, using one of many routing protocols, to gain information about the network status and build their routing tables. Therefore, routing tables are automatically updated as the network changes in the dynamic routing case.

4.2.3 Use of ICMP messages in routing

ICMP redirect

When enabled, a router sends an ICMP redirect error message to the sender of a datagram if the datagram should have been sent to another router. This allows the host to build a better routing table. The host may start with just a default router in its routing table. ICMP redirects from the default router will allow the host to update and build its routing table.

Figure 4.1 gives the format of an ICMP redirect message. Figure 4.2 shows an ICMP redirect example, where Host X uses Router A as its default router. When Host X has a datagram to send to Host Y, it sends the datagram to its default router. When Router A receives the datagram, it

0	7 8	15 16	31
type(5)	code(0-3)	checksum	
correct gateway IP address			
IP header (including options), plus the first 8 bytes of the original IP datagram payload			

Figure 4.1. ICMP redirect message format.

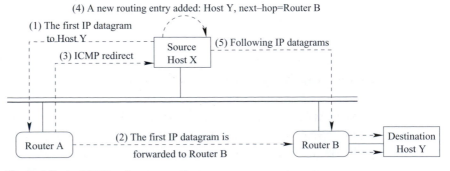

Figure 4.2. An ICMP redirect example.

looks up the routing table and decides that Router B is the next-hop router. However, it detects that the datagram is being sent out on the same interface it was received on. Then, Router A sends an ICMP redirect message to Host X, saying that subsequent datagrams to Host Y should be routed to Router B. After receiving the ICMP redirect message, Host X inserts a more efficient routing entry for Host Y using Router B as the next-hop router, with a D flag.

ICMP redirect is enabled by default in the routers in our lab. To enable this feature if it is disabled, use the following command in the router *Interface Configuration* mode:

> Router(config-if)# **ip redirects**.

ICMP router discovery

In networks of moderate size and simple topology ICMP router solicitation and ICMP router advertisement messages can be used to configure the default route for a host when it bootstraps. When a host boots up, it sends several ICMP router solicitation messages, a few seconds apart, to the multicast IP address 224.0.0.2 (ALL_ROUTERS). A router interface listens on the ALL_ROUTERS address and responds to ICMP router solicitation messages with ICMP router advertisement messages. In addition, the router also sends ICMP router advertisement messages periodically. A host chooses one or more of the advertised addresses as its default router. This process is called *ICMP router discovery*.

In Red Hat Linux, router solicitations and router advertisements work only on IPv6 networks.

In Solaris, the /etc/defaultrouter file stores the default router (IP or domain name). If this file is empty, the host will use ICMP router discovery to find a default route. The routing daemon for ICMP router discovery is /usr/bin/in.rdisc.

To enable the ICMP router discovery protocol on a router interface, type the following in the *Interface Configuration* mode:

> Router(config-if)# **ip irdp**.

An ICMP router solicitation message is 8 bytes long, with the 1-byte Type set to 10, the 1-byte code set to 0, and a 2-byte checksum. The remaining 4 bytes are unused and set to all 0s. An ICMP router advertisement message carries one or more router IP addresses, as shown in Fig. 4.3.

0	7 8	15 16	31
type(9)	code(0)	checksum	
no. of addr.	addr. length(2)	lifetime	
router address [1]			
preference level [1]			
router address [2]			
preference level [2]			
......			

Figure 4.3. ICMP router advertisement message format.

4.2.4 Dynamic routing

The Internet is a collection of networks with a very complex and time-varying topology. Dynamic routing is needed to eliminate loops in the paths and to react to changes in the network topology (e.g., a link failure).

The Internet is organized as a collection of *Autonomous Systems* (AS). An AS is a set of networks administered by a single entity, e.g., an enterprise network, or a campus network. Routing protocols used inside an AS are called *interior gateway protocols* (IGP), while those used between routers in different ASs to interconnect them are called *exterior gateway protocols* (EGP). The most widely used EGP protocol is the Border Gateway Protocol (BGP). Classless Interdomain Routing (CIDR) is used to reduce the size of the Internet routing tables. We will focus on IGPs in this chapter and examine two popular IGPs: the *Routing Information Protocol* (RIP) and the *Open Shortest Path First* protocol (OSPF). Then, we will discuss CIDR briefly.

Link state and distance vector routing

Routing algorithms are the core of dynamic routing protocols. These algorithms use a metric to determine the optimal path to a destination. Different metrics can be used, such as path length, reliability, delay, bandwidth, load, and communication cost.

Generally there are two types of routing algorithms, namely, *distance vector routing* and *link state routing*. A distance vector routing algorithm exchanges all or a portion of the router's routing table with its neighbors in terms of a vector of distances (number of hops) to destination networks. The routing table is then computed using the routing information (destination and distance pairs) received from its neighbors. A router running a link

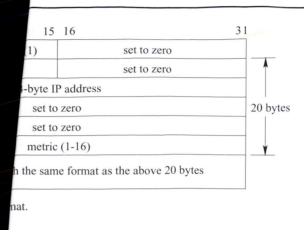

15	16		31	
(1)	set to zero			
	set to zero			20 bytes
-byte IP address				
set to zero				
set to zero				
metric (1-16)				
h the same format as the above 20 bytes				

mat.

15	16		31	
sion(2)	routing domain			
	route tag			20 bytes
4-byte IP address				
4-byte subnet mask				
-byte next-hop IP address				
metric (1-16)				
, with the same format as the above 20 bytes				

ge format.

imer. Amount of time that must pass before the route
the routing table. The interval should be longer than
invalid and hold-down values. The default value is 240

re encapsulated in UDP datagrams, using the well-known
Figure 4.4 shows the format of a RIP message, and Fig. 4.5
of a RIP-2 message. The fields of a RIP message are listed

ates whether the packet is a request (1) or a response (2).
ber: Specifies the RIP version used (1 or 2).
ily Identifier: Specifies the address family used. RIP
carry routing information for several different protocol
this field is 2.
es the IP address for the entry.
how many hops have been traversed from the source

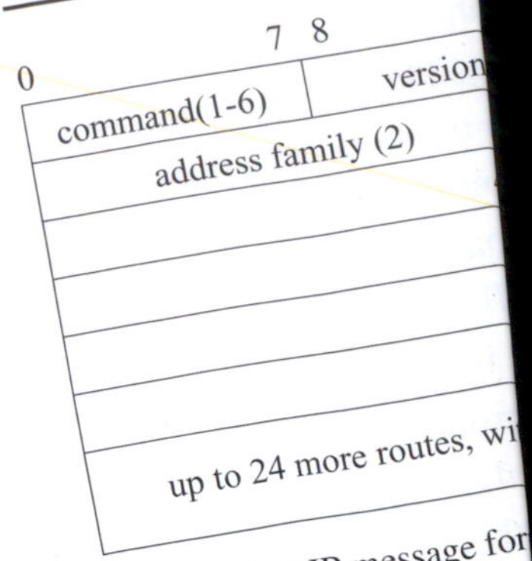

Figure 4.4. RIP message for...

RIP

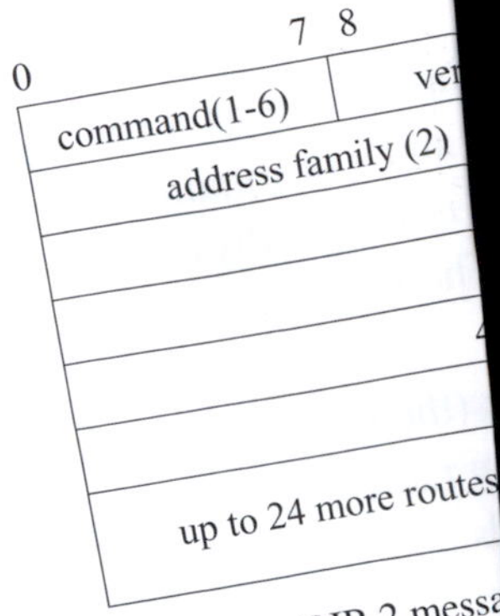

Figure 4.5. RIP-2 messa...

4. The *route-flush* t...
   is removed from...
   the larger of the...
   seconds.

   RIP messages ar...
   port number 520. ...
   shows the format ...
   here.

   • Command: Indic...
   • Version Num...
   • Address-Fa...
     can be used...
     families. F...
   • Address...
   • Metri...
     to the...

1

2.

   (
   a
   tc
   is

3. Th
   tior
   thre
   dow
   unrea

[1] Refer to t...

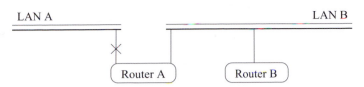

Figure 4.6. Illustration of the Count-to-Infinity problem.

The RIP-2 message takes advantage of the unused fields in RIP, and provides additional information such as subnet support and a simple authentication scheme. These fields are listed here.

- `Routing Domain`: The identifier of the routing daemon that sends this message (e.g., the process ID of the routing daemon).
- `Route Tag`: Used to support EGPs, carrying the AS number.
- `Subnet Mask`: The subnet mask associated with the IP address advertised.
- `Next-hop IP Address`: Where IP datagrams to the advertised IP address should be forwarded to.

RIP is widely used because of its simplicity and low routing overhead. However, it has the *Count-to-Infinity* problem which causes routing loops. Consider the network in Fig. 4.6. Initially `Router A` connects to `LAN A`, and `Router B` has a routing entry that shows the route to `LAN A` is through `Router A` in one hop. When the link to `LAN A` fails, `Router A` examines the RIP message from `Router B` and sees that `Router B` has a one-hop route to `LAN A`. Then, `Router A` advertises a two-hop path to `LAN A` and routes all `LAN A` bound traffic to `Router B`, resulting in a loop. When `Router B` sees the advertisement from `Router A`, it changes its entry for `LAN A` to a three-hop route through `Router A`. This process will continue indefinitely, with the route lengths in both routing tables increasing to infinity.

To solve this problem, RIP uses a *hop-count limit* of 15. In the above example, when the path length in either routing tables reaches 16, `LAN A` will be regarded as unreachable. In the above example, the loop will be eliminated when the hop-count limit is reached. The downside of this hop-count limit approach is that the size of the network running RIP is limited. It also takes a long period for the routing tables to converge after a topology change. In addition to the hop-count limit, routers hold down any changes that might affect recently removed routes for the hold-down timer interval, in order to let the failure propagate through the entire network. Another technique to improve the stability of RIP is *split horizon*, where information about a route is not allowed to be sent back in the direction from which it came.

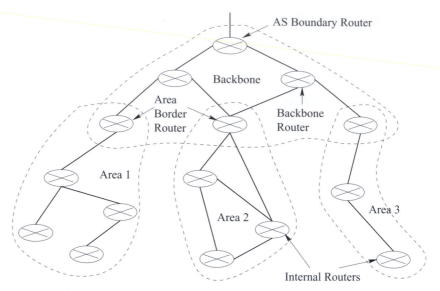

Figure 4.7. An AS with a backbone and three areas.

Open shortest path first

OSPF was standardized in the mid-1980s to overcome the limitations of RIP. OSPF is a link state routing protocol, where each OSPF router sends *link state advertisements* (LSA) to all other routers within the same hierarchical area. LSAs are messages that include information on attached interfaces, metrics used, and other variables. OSPF routers accumulate link state information in their link state databases to learn the network topology and use the *shortest path first* (SPF) algorithm to calculate the shortest path to each node. The `gated` routing doemon uses RIP, RIP-2, and OSPF.

OSPF operates within a hierarchy, which is different from RIP. An autonomous system (AS) is the largest entity within the hierarchy. Then, the AS is partitioned into several *areas*. The topology of an area is invisible to entities outside this area, so that OSPF passes less routing traffic than it would without the partitioning of the AS. An OSPF backbone, consisting of all area border routers, networks not completely contained in any area, and their attached routers, distributes routing information between areas. Figure 4.7 shows an AS with a backbone and three areas.

After a router is assured that its interfaces are functioning, it sends *OSPF Hello* packets to its neighbors. The router also receives hello packets from its neighbors. In addition to helping detect neighbors, hello packets also let routers know that other routers are functional. Two routers are said to

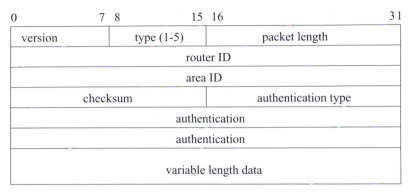

Figure 4.8. OSPF message format.

be *adjacent* when their link state databases are synchronized. Each router periodically sends LSAs to provide information on the link states, so that failed routers can be detected quickly. By using the information in LSAs, a router builds a *topological database* containing an overall picture of the area, and calculates a shortest-path tree with itself as root, which then yields a routing table.

Rather than using TCP or UDP, OSPF uses IP directly. In the IP header, OSPF has its own value for the `protocol` field (89, see Fig. 0.7). Each OSPF packet has a 24-byte header, as shown in Fig. 4.8. The fields in an OSPF packet header are given.

- `Version Number`: Specifies the OSPF version used.
- `Type`: Identifies the OSPF packet type as one of the following.
  - `Hello`: Establishes and maintains neighbor relationships.
  - `Database Description`: Describes the topological database content.
  - `Link-state Request`: A request for the topological database from neighbor routers when a router discovers that parts of its topological database are out of date.
  - `Link-state Update`: A response to a link-state request packet.
  - `Link-state Acknowledgement`: Acknowledges link-state update packets.
- `Packet Length`: Specifies the packet length in bytes, including the OSPF header. The maximum length is $2^{16} - 1 = 65535$ bytes since this field is 16 bits long.
- `Router ID`: Identifies the source of the packet.
- `Area ID`: Identifies the area to which the packet belongs.
- `Checksum`: Checks the entire packet contents for transmission errors.

- `Authentication Type`: Specifies the authentication type. All OSPF protocol exchanges are authenticated. This field is configurable on a per-area basis.
- `Authentication`: Contains authentication information.

Classless interdomain routing (CIDR)

With the exponential growth of the Internet, routing tables in the core routers are getting longer and longer. For example, there needs to be a routing entry for each newly assigned Class C network, and there could be as many as 2^{21} Class C networks. With current technologies, it is impossible to have such a huge routing table in a core router. To further reduce the routing table sizes, CIDR uses a technique called *supernetting* to summarize multiple routing entries into a smaller number of routing entries.

With CIDR, IP addresses are not classified into classes anymore ("classless addressing"). Rather, each IP address consists of two components: a network prefix ranging from 13 to 27 bits, and a host ID using the remaining bits. The *slash notation* is used to denote an IP address, as "A dotted-decimal IP address" + "/" + "Number of bits used for the network prefix". For example, the slash notation 128.238.66.100/24 means the network prefix has 24 bits, i.e., 128.238.66. In CIDR, the network addresses are assigned in a hierarchical manner. For example, an ISP is assigned a network address with a shorter prefix (e.g., 128.238/16), while each client network of the ISP is assigned the same network address but with a longer prefix (e.g., 128.238.61/24 and 128.238.62/24). In the core network, routing entries for a number of networks with the same higher level prefix can be summarized into a single entry. Therefore, the routing table size can be greatly reduced. The *longest-prefix-matching* rule is used in table lookups.

The Internet is currently a mixture of both CIDR-type addresses and the traditional Class A, B, and C addresses. CIDR is supported in almost all new Internet authorities strongly encourage its deployment.

4.2.5 Multiple protocol label switching and traffic engineering

Multi-protocol label switching (MPLS) is a virtual-circuit packet switching technology overlayed on the datagram (connectionless) packet-switched Internet. This is achieved by attaching short fixed length *labels* to packets. The MPLS label is located after the layer two header and before the IP header. In an MPLS network, labels attached to packets are used to make forwarding decisions. *Label switching routers* can make forwarding decisions based on the label contents rather than the more time consuming table lookups based on IP addresses.

In a large network, it is possible that available network bandwidth is not efficiently utilized because the intra-domain routing protocol, such as OSPF, finds path based on a single "least-cost" scalar metric for each destination. This least cost route may not have enough resources to carry all the traffic, or satisfy all the service requirements of carried traffic. Congestion, at certain hot spots, can result in sub-optimal use of network resources. In MPLS networks *label switched paths* (LSPs) can be set up in MPLS networks to route around congestion and to create paths for each traffic type, e.g., Voice over IP (VoIP) or best-effort data. This allows service provider to use *traffic engineering* (TE) techniques to maximize the utilization of network resources, and/or enhance the quality of service it can offer. LSPs also allow network service providers to set up IP tunnels for *virtual private networks* (VPNs). VPN customers, typically large corporations, get the security and performance benefits of a private data network by setting up IP tunnels using LSPs in the service provider's MPLS network.

4.3 Manipulating routing tables

4.3.1 Routing table for a workstation

To build the routing table in a workstation, the **route** command is used. Routing entries that are created manually usually do not change, even if you run a routing daemon. Static routes are often added from the system startup script. The following commands can be used to display or build the host routing table.

- **netstat -rn**: Displays the routing table in the system.
- **route   add   [-host|-net]** *destination_address*[/*prefix*] **[gw** *gateway_address*] **[metric** *M*] **[netmask** *mask*] **dev** *interface_name*: Inserts a new route to the routing table. *routing_metric* is the hop count to *destination_address*.
- **route   del   [-host|-net]** *destination_address*[/*prefix*] **[gw** *gateway_address*] **[metric** *M*] **[netmask** *mask*] **dev** *interface_name*: Deletes an existing route from the routing table.
- **route add|del default gw** *router_interface interface_name*: Adds (deletes) a default route entry.

In Solaris 8.0, the **route** commands are:

- **route add [net]** *dest_addr gateway_addr metric*: Inserts a new route to the routing table.
- **route delete [net]** *dest_addr gateway_addr*: Deletes an existing route from the routing table.
- **route add|delete default** *router_addr* **1**: Adds (deletes) a default routing entry.

4.3.2 Routing table for a router

You can also manually configure the routing table in a router. To display the router routing table, use the following command in the *Privileged EXEC* mode:

router# **show ip route**.

The following *Global Configuration* commands can be used to modify the router routing table.

- Router(config)# **ip route** *prefix mask next_hop* [*admin_distance*]: Creates a route to the destination subnet with IP prefix *prefix* and subnet mask *mask*, via *next_hop*. The *admin_distance* argument is optional. If you want a static route to be overridden by dynamic routing information, specify *admin_distance* greater than the default administrative distance of the routing protocol. As an example, a RIP-derived route has the default value of 120.
- Router(config)# **no ip route** *destination next_hop*: Removes a route from the IP routing table.

4.4 Traceroute

Traceroute is a tool that helps determine all the routers in an end-to-end path. It uses the Time-to-Live (TTL) field in the IP header and the ICMP protocol. TTL is an 8-bit field usually set to 64. Each router that sees the datagram decreases the value of TTL by one. If a router receives a datagram with a TTL value of 1, it discards the datagram and sends an ICMP time exceeded message back to the source.

Traceroute works as illustrated in Fig. 4.9. First, it sends an IP datagram to the destination host with the TTL field set to 1. The first router seeing

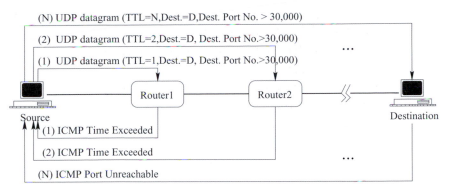

Figure 4.9. The operation of Traceroute.

the datagram decrements the TTL, discards the datagram, and returns an ICMP `time exceeded` message to the sender. The datagram carrying this ICMP message also contains the router's IP address as the source address. Thus the first router in the path is identified. Next, Traceroute sends a datagram with a TTL of 2, and the address of the second router is identified in a similar way. This continues until the destination host is reached. The destination host will not discard the datagram even though the TTL field is 1 (because the packet has reached its destination). To find out whether the destination host is reached, Traceroute chooses a large UDP destination port number (greater than 30,000), which is unlikely to be in use by any process at the destination host. Then the destination will return an ICMP `port unreachable` message to the source. In summary, the ICMP `time exceeded` messages identify the intermediate routers, and the ICMP `port unreachable` message identifies the end host.

4.5 A simple router experiment

As in the previous lab, we will divide the students into four groups, each with two workstations, a router, and two hubs, which are to be connected as shown in Fig. 4.10. The IP addresses of the routers and hosts are given in Table 4.2 and Table 4.3, respectively.

Exercise 1   Configure the IP addresses of your workstations and the router as shown in Fig. 4.10, Table 4.2 and Table 4.3.

Initially your host's routing table has no entry for the subnet on the other side of the router. In order to be connected, you need to add a routing entry for the other subnet in the routing table of your workstation (see section 4.3.1).

Table 4.2. *Router IP addresses for Fig. 4.10*

	eth0	eth1
router1	128.238.61.1/24	128.238.62.1/24
router2	128.238.62.2/24	128.238.63.2/24
router3	128.238.63.3/24	128.238.64.3/24
router4	128.238.64.4/24	128.238.65.4/24

Table 4.3. *Host IP addresses for Fig. 4.10*

HOST_A			HOST_B		
Name	IP address	Label	Name	IP address	Label
shakti	128.238.61.101/24	1	vayu	128.238.62.101/24	2
agni	128.238.62.102/24	3	apah	128.238.63.102/24	4
yachi	128.238.63.103/24	5	fenchi	128.238.64.103/24	6
kenchi	128.238.64.104/24	7	guchi	128.238.65.104/24	8

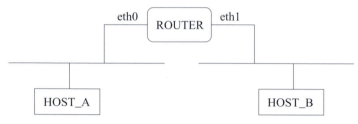

Figure 4.10. Simple router experiment.

Run **tcpdump -en** on your machine, and **tcpdump -en** on your partner's machine in the other subnet simultaneously:

> **tcpdump -en host** *remote_host* **and** *your_machine.*

Send **ping** messages continuously to your partner's machine:

> **ping -sv** *remote_host.*

After receiving the tenth echo reply, quit **ping** and save the **tcpdump** outputs from both machines. Also, copy the **ping** output.

During this exercise, don't run the **ping** program at the same time. For clean results, do your experiments in turn.

LAB REPORT   When a packet was sent to a workstation in the other subnet, explain how the source and destination Ethernet addresses were changed.

What are the source and destination addresses in the IP and Ethernet headers of a packet that went from your machine to the router?

What are the source and destination addresses in the IP and Ethernet headers of a packet that went from the router to your partner's machine?

Answer the above two questions, but now for the echo reply that was returned from your partner's machine.

LAB REPORT   Use the **tcpdump** outputs from both machines to calculate the average delay that a packet experienced in the router. Note that the system times of the two machines might be different. Show all the steps and submit the **tcpdump** outputs with your report.

Compare this value with the previous value in the case of the bridge. Which, a router or a bridge, is faster? Why?

4.6 RIP exercises

In this section, we will examine the operation of RIP. To enable the RIP routing process in a router, use the following commands in the *Global Configuration* mode.

Router(config)# **router rip**

Router(config)# **network** *network_number,*

where *network_number* could be 128.238.0.0. To remove the network, use:

Router(config)# **no network** *network_number*

To shutdown the RIP process, use:

Router(config)# **no router rip**

Consider Fig. 4.11 as our network topology for this section. Since the IP address of ethernet1 in router4 is the only interface which is different from the initial configuration in Appendix B, we will reboot all the four routers to restore their default configurations, and change the IP address on the ethernet1 in router4 only. Since our workstations started **routed** at boot-up time, no further action is needed to run RIP on the workstations.[2]

Exercise 2   Connect the routers and hosts and change the IP addresses of the workstations and router4 as shown in Fig. 4.11. Also, make sure that your workstation has no other routing entries than your own subnet and your loopback interface. For how to remove an entry from the host routing table, see Section 4.3.

Run the RIP process in each router. To avoid confusion, each router should be configured by only one person.

---

[2] The lab instructor should make sure that the IP-Forwarding function is enabled in each host (see Appendix A.6).

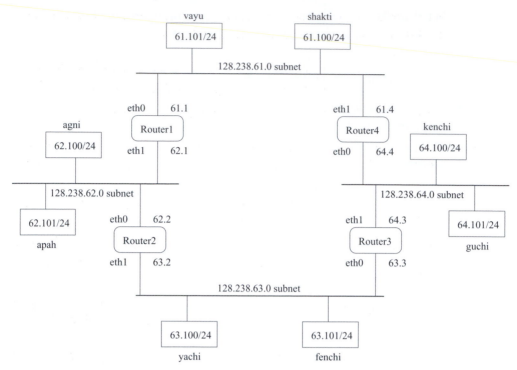

Figure 4.11. Network configuration of the RIP experiment.

After starting RIP in all the routers, test connections to other hosts by pinging them. Once you can successfully reach all the hosts, run the following command to capture the RIP messages sent on your subnet:

tcpdump -x -s 100 -c 4 -w exe2.out udp port 520.

Save the routing table in your workstation. Note the number of hops needed to reach destinations other than in your own subnet.

Run **ethereal -r exe2.out &** to load the packet trace recorded by the above **tcpdump** command. Mark and print two different RIP messages captured in your subnet (see Exercise 6 of Chapter 1). Exchange the printed RIP messages with students in other groups. You need eight different RIP messages for your lab report.

LAB REPORT　Explain why you can only get two different RIP messages in your subnet. Was a RIP packet forwarded by the routers? Why?

LAB REPORT　Draw the format of one of the saved RIP response packets from your subnet, including the IP and UDP headers and the RIP message (see Figs 0.13, 0.14, and 4.4). Identify each field, and express their values in decimal format.

For the other seven RIP response packets collected, explain the contents of the RIP messages only, excluding IP and UDP headers.

LAB REPORT   Draw the distance tables and the routing tables in the routers based on Fig. 4.11, assuming that number of hops is used as the metric.

Verify the routing tables using the RIP messages you captured.

Exercise 3   In this exercise we will examine how RIP responds to link failures. Send **ping** message continuously from `apah` to `yachi` and start **tcpdump** on `apah`. Let the two programs run during this exercise.

Disconnect the cable from the `ethernet0` port of `router2` from the hub in the 128.238.62.0 subnet, and type the **time** command to get the current time.

Observe the **ping** and **tcpdump** windows. When the connection is re-established, type the **time** command again. See how much time RIP takes to alter the routing table in your workstation to the new topology.

Once you can successfully reach other hosts, connect the cable to the original position. Again, measure the time that RIP takes to change your routing table.

LAB REPORT   Compare this time with the previous value in the spanning tree experiment.

Explain why it takes this time for RIP to react to the route change. Refer to Section 4.2.4 for RIP operation and default timer values.

4.7 Routing experiments with ICMP

Exercise 4   Eliminate the routing entries for subnets other than your own and the loopback interface. Save the routing table for your lab report.

Create a default routing entry using one of the routers directly connected to your workstation.

While **tcpdump -enx -s 100 ip proto 1** is running, send **ping** messages to a host that is three hops away through the default router.

After capturing an ICMP redirect message, save the **tcpdump** output, the **ping** output, and your workstation's routing table. You may need to **ping** the same host several times in order to get your routing table updated.

LAB REPORT   Submit what you saved in Exercise 4.

Identify every field in the ICMP redirect message (see Fig. 4.2).

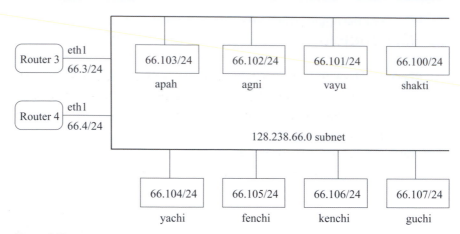

Figure 4.12. Network configuration of the ICMP router discovery experiment.

Compare the original routing table with the new routing table. Explain the meaning of the flags of the new entry.

Exercise 5 This exercise[3] is on ICMP router discovery. All students should do this exercise together, using a single segment network.

Connect the routers and hosts and change the host IP addresses as shown in Fig. 4.12.

Telnet to the routers, change the IP address of the `ethernet1` interfaces as shown in Fig. 4.12. Enable ICMP router discovery on these two interfaces by the following *Interface Configuration* command:

> `Router(config-if)`# **ip irdp**.

Run **tcpdump -enx ip proto 1** on all the hosts except `shakti`.

The lab instructor should now reboot `shakti`.

Save the captured route discovery requests and replies for the lab report.

Telnet to `shakti` and save its routing table for the lab report.

LAB REPORT   What is the destination IP address of the ICMP router solicitation message? Who sends the ICMP router advertisement message?

What are the `type` and `code` of the ICMP messages captured?

What are the advertised router IP addresses and their preference levels?

How many default router entries are there in `shakti`'s routing table? Why?

---

[3] This exercise is for Solaris only, since Red Hat Linux does not support ICMP router discovery.

4.8 OSPF exercise

In order to enable OSPF in the routers, you need to create an OSPF routing process first. Then, define the range of IP addresses to be associated with the routing process and assign area IDs for these IP addresses, using the following commands:

`Router(config)#` **router ospf** *process_id*

`Router(config)#` **network** *address wildcard_mask* **area** *area_id*.

Process_id is a numeric value local to the router. It does not have to match process_ids on other routers. *Address* is the network address of the interface on which the OSPF process runs (128.238.0.0 in our case). *Wildcard_mask* helps reduce the number of configuration commands. 0 is a match and 1 is a "don't care" bit (0.0.255.255 in our case). *Area_id* is the number of the area that the interfaces belong to (see Fig. 4.7). It can be any integer between 0 and $2^{32} - 1$ or can have an IP address form. Note that 0 is reserved for the backbone.

The above commands are required to configure OSPF, while other tasks (configuring interface parameters, configuring area parameters, etc.) are optional. For more information on other configuration tasks, refer to the router manual.

Consider Fig. 4.13 for our OSPF experiment. The lab instructor should reboot the routers to restore their default configurations.

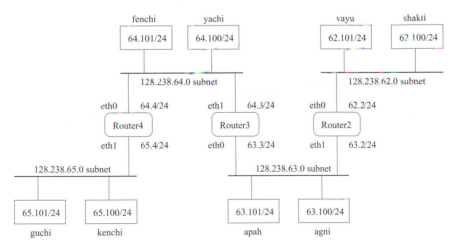

Figure 4.13. Network configuration for the OSPF exercise, the static routing exercise, and the **traceroute** exercise.

Exercise 6   After connecting the cables properly, change the host IP addresses as given in Fig. 4.13. You need to remove the default route added in Exercise 4 from the host routing table. Note that the router interfaces are set as Fig. 4.13 by default.

Run the following command to capture any OSPF packets:

tcpdump -x -s 120 ip proto 89

Login to a directly connected router and start the OSPF process. Set the argument *area_id* to 1 for all the routers.

The workstations in our lab run **routed** (which uses RIP). The routing daemon supporting OSPF, **gated**, is not installed. In order to reach the routers and hosts in the other subnets, you need to add a default router in your host's routing table.

Examine the routing table in each router (see Section 4.3). When the routing table gets an entry for the network that is not directly connected, kill the **tcpdump** process and save the **tcpdump** output.

Collect the **tcpdump** outputs from other subnets. Study the various types of OSPF packets from the **tcpdump** outputs.

You can display OSPF information in a router using the following commands in the *Privileged EXEC* mode.

```
show ip ospf
show ip ospf database [router|network|summary| \
                  asb-summary|external|database-summary]
show ip ospf interface ethernet [0|1]
show ip ospf neighbor
show ip ospf virtual-links
```

LAB REPORT   Draw the common header of a saved OSPF message, giving the decimal values of the header fields (see Fig. 4.8).

Submit the routing tables you collected from the routers.

4.9 Static routing experiment

In this experiment, we reuse the network as shown in Fig. 4.13.

Exercise 7   After checking the wiring, as shown in Fig. 4.13, reboot the routers to restore their initial settings. Check the IP addresses of the workstations as shown in Fig. 4.13.

Remove all the routing entries other than your own subnet and the loopback interface from your host routing table. Save the output of **netstat -rn** before building your workstation's routing table.

Examine Fig. 4.13 and build your host's static routing table manually.

Telnet to a router that is directly connected to your workstation, and save its routing table before building any route. Save the routing table of the other router if you have one more router connected directly. You may not be able to **telnet** to a router that is not directly connected. In this case, copy the initial routing table of these routers from students in other subnets later.

Now configure the routing table in each router. See Section 4.3 for commands and syntax on manipulating router routing tables. Note that each router should be configured by one person only.

Use **ping** to test the connections. When you can reach all other subnets successfully,[4] save the routing tables in your workstation and all the routers for the lab report.

LAB REPORT   Submit the routing tables saved in this exercise.

4.10 Traceroute experiment

In this exercise, we use the same network and configuration of the previous exercises, and use **traceroute** to find a multi-hop path.

| Exercise 8 | Execute **tcpdump -enx -s 100 host** *your_host* **and** *remote_host* on your host, where *remote_host* is a workstation at least two hops away.

Then, execute **traceroute** *remote_host* to find the route from your host to the remote host.

Save the output of both **traceroute** and **tcpdump**.

LAB REPORT   Submit what you saved in this exercise.

From the **tcpdump** output, explain how the multi-hop route was found. Explain the sequence of the ICMP messages used.

---

[4] Even when the routing table in your workstation and all the routers are configured perfectly, you may not be able to **ping** a remote host, if the routing table in the remote host is incorrect. When you can get **ping** reply messages from all the interfaces of the routers successfully, your work is done for this exercise.

5   UDP and its applications

The principle, called the end-to-end argument, suggests that functions placed at low levels of a system may be redundant or of little value when compared with the cost of providing them at that low level.

J. H. Saltzer, D. P. Reed and D. D. Clark

5.1 Objectives

- Study **sock** as a traffic generator, in terms of its features and command line options.
- Study the User Datagram Protocol.
- IP fragmentation.
- MTU and path MTU discovery.
- UDP applications, using the Trivial File Transfer Protocol as an example.
- Compare UDP with TCP, using TFTP and the File Transfer Protocol.

5.2 The User Datagram Protocol

Since the Internet protocol suite is often referred to as TCP/IP, UDP, it may seem, suffers from being considered the "less important" transport protocol. This perception is changing rapidly as realtime services, such as Voice over IP (VoIP), which use UDP become an important part of the Internet landscape. This emerging UDP application will be further explored in Chapter 7.

UDP provides a means of multiplexing and demultiplexing for user processes, using UDP port numbers. It extends the *host-to-host* delivery service of IP to the *application-to-application* level. There is no other transport control mechanism provided by UDP, except a checksum which protects the UDP header (see Fig. 0.14), UDP data, and several IP header fields.

Compared with the other transport protocol, TCP, UDP is simpler in the sense that it does not guarantee successful and in-order delivery of the datagrams. UDP is used by many network services, such as DNS, TFTP (which we will examine in this chapter), NFS, RPC, BOOTP/DHCP, and SNMP. UDP is also suitable for realtime services, such as video streaming and VoIP, which are delay sensitive and loss tolerant. Besides *unicast* service, UDP also provides *multicast* service. We will examine UDP multicast and realtime transport in Chapter 7.

5.3 MTU and IP fragmentation

5.3.1 IP fragmentation

In Chapter 2 we saw that an important parameter associated with each network interface is the MTU. An interesting question with MTU is what happens if an IP datagram is longer than the MTU of the interface. In this case, the IP layer splits the datagram into several fragments, each with a length less than or equal to the MTU. This process is called *IP fragmentation*.

The following IP header fields (see Fig. 0.13) are related to IP fragmentation.

- `Total Length`: After fragmentation, this changes to the size of the fragment in the IP datagram.
- `Identification`: All fragments from the same IP datagram carry the original Identification.
- `Flags`: The "more fragments" flag indicates if the current fragment is the last one or not, while the "don't fragment" flag can be set by the source to disallow fragmentation in intermediate routers.
- `Fragment Offset`: contains the offset (in 8-byte units) of the current fragment in the original datagram.

Fragmentation may occur in the source host or an intermediate router. A fragment may be further fragmented if it is sent to a link with a smaller MTU. However, reassembly of the fragments is only performed at the receiver, where IP datagrams with the same identification are put together to reconstruct the original IP datagram. The `More Fragments` flag and the `Fragment Offset` field are used to put the fragments in the right order. If fragmentation is needed but the `Don't Fragment` flag is set, the router drops the datagram and returns an ICMP unreachable error message back to

0	7 8	15 16	31
type (3)	code (4)	checksum	
unused (set to 0)		MTU of the next–hop network	
IP header (including options) + first 8 bytes of the original IP datagram data			

Figure 5.1. The format of an ICMP unreachable error message.

the source. This feature is used in path MTU discovery to find the smallest MTU along a path.

5.3.2 Path MTU discovery

The minimum MTU of the links along a path is called the *path MTU*. Since fragmentation degrades router performance, a source host can perform *path MTU discovery* to find the path MTU. It can then send datagrams no longer than the path MTU to avoid fragmentation in the routers.

In path MTU discovery, a host sends IP datagrams with the "don't fragment" bit set. If the MTU of a link is smaller than the IP datagram, the router drops the datagram and send an ICMP unreachable error to the source carrying the MTU of the next link. Figure 5.1 shows the format of an ICMP unreachable error, where byte 7 and 8 store the MTU of the next-hop network.

The following *Interface Configuration* command enables the router to send ICMP unreachable errors:

Router(config-if)# **ip unreachables**.

The MTU of a host interface can be modified using the following command:

ifconfig *interface_name* **mtu** *new_MTU_value*.

To set the MTU of a router interface, use the following *Interface Configuration* command:

Router(config-if)# **ip mtu** *new_MTU_value*.

5.4 Client–server applications

5.4.1 The client–server architecture

Most network applications are implemented using a client–server architecture, where a server provides network service to the clients. Servers use well-known port numbers (defined in the /etc/services file), and are usually running all the time, whereas a client uses an ephemeral port number and terminates after receiving the service. If a client requests a service on a port number that no server is associated with, an ICMP port unreachable error is returned to the client in the case of a UDP packet, and the TCP connection is reset if TCP is used. In the following, we discuss two application layer protocols that provide file transfer service.

5.4.2 TFTP

TFTP is a simple file transfer protocol using UDP. Since UDP is connectionless and unreliable, TFTP uses a *stop-and-wait* flow control algorithm, where each data packet is acknowledged by an ACK packet before the next data packet is sent. In addition, a lost packet causes timeout and retransmission. TFTP is primarily designed for diskless systems to download configuration files from a remote server during bootstrapping.

Figure 5.2 shows the architecture of a TFTP session. A common feature for all the application layer protocols is the user interface (UI) module. A UI directly interacts with a user, by translating user inputs (such as keyboard entries and mouse clicks) into protocol primitives and displaying the results of the operations. The TFTP protocol interpreter accesses the local file system and communicates with its counterpart at the other end of the session. The TFTP server uses UDP port 69 for TFTP control messages. A different ephemeral port number is used by the server for data transfer.

Figure 5.3 shows the packet format of TFTP messages. The *opcode* field is used to multiplex different TFTP messages. A typical TFTP session, where a client downloads a file from the server, is as follows.

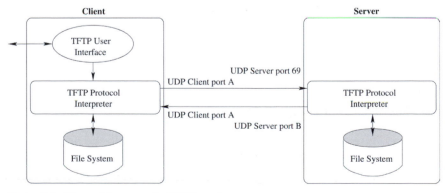

Figure 5.2. The architecture of a TFTP session.

opcode (1=RRQ, 2=WRQ)	filename	0	mode	0
2 bytes	variable length	1 byte	variable length	1 byte

opcode (3=data)	block number	data
	2 bytes	0~512 bytes

opcode (4=ACK)	block number

opcode (5=error)	block number	error message	0
		variable length	1 byte

Figure 5.3. The TFTP packet formats.

1. A client sends a read request (RRQ) to a server on UDP port 69.
2. The server responds with data packets (if the requested file exists) of length 512 bytes and block number starting with 1.
3. The client sends an acknowledgement for the received block.
4. The server sends the next block with the block number increased.
5. The above two steps continue until the last block which is shorter than 512 bytes is received.

As its name implies, TFTP is designed for small and infrequent file transfers, where throughput is not a major concern. In other cases where bulk data transfer is needed, FTP, using TCP's window flow control, is used for better throughput performance. Another limitation of TFTP is the lack of a login procedure. This is a "security hole" in TFTP.

5.4.3 FTP

FTP is a file transfer protocol using TCP. Figure 5.4 shows the FTP architecture, where two TCP connections are used: a control connection (TCP port 21) for FTP commands and replies, and a data connection (TCP port 20) for file transfer.

To set up an FTP session, the client sends SYN request to the server TCP port 21 to establish the control connection. TCP connections will be discussed in the next chapter. Tables 5.1 and 5.2 give the FTP commands and typical server replies that can be sent on the control connection. User inputs (e.g., **get foo.txt**) are translated to the primitives (e.g., **RETR foo.txt**) shown in Table 5.1 by the UI, and sent on the control connection. In addition, server responses, shown in Table 5.2, are received from the control connection and are translated to more friendly messages by the UI

Table 5.1. *Common FTP commands*

Command	Description
List *field*	list files or directories
PASS *password*	password on server
PORT *n1,n2,n3,n4,n5,n6*	client IP address (*n1.n2.n3.n4*) and port ($n5 \times 256 + n6$)
QUIT	log off from server
RETR *filename*	retrieve (get) a file
STOR *filename*	store (put) a file
TYPE *type*	specify file type: A for ASCII, I for image
USER *username*	username on server

Table 5.2. *Typical FTP replies*

Reply	Description
125	Data connection already open; transfer starting.
200	Command OK.
331	Username OK, password required.
425	Can't open data connection.
452	Error writing file.
500	Syntax error (unrecognized command).
501	Syntax error (invalid arguments).

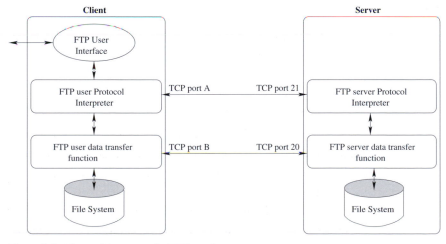

Figure 5.4. The architecture of a FTP session.

and displayed on the client screen. A data connection is created each time a file is transferred. To open a data connection, the client first chooses an ephemeral port number and then sends the port number to the server using the PORT command, as shown in Table 5.1, via the control connection. Then the server issues an active open to that port on the client host. File transfer begins after the data connection is set up.

Many FTP servers support *Anonymous FTP*, which allows everyone to log in and perform file uploads and downloads. Public domain free information is sometimes provided using this technique. The login name of Anonymous FTP is anonymous, and the password is your own email address.

Most FTP implementations can be run in the *debug* mode, which is a convenient way to study the operations of FTP. To run **ftp** in the debug mode, use:

ftp -d *ftp_server_IP*.

5.5 Using the sock program

Exercise 1   Use the following commands to observe the basic operation of **sock**.
- **sock** *host* **echo**
- **sock -s 5555**
- **sock -i -n3 -w2048** *host* **5555**

LAB REPORT   Explain the operation of each command.

Exercise 2   Study various options associated with the **sock** program. A brief list of options can be displayed by typing **sock**. More detailed discussion on **sock** can be found in Appendix C of [5].

5.6 UDP exercises

Exercise 3   While running **tcpdump src host** *your_host*, execute the following command with different values of *size* (i.e., the size of the datagram).

sock -u -i -n1 -w*size remote_host* **echo**

The **-u** option is used to send UDP datagrams rather than TCP segments.

Increase *size* (i.e. the size of the datagram) until fragmentation occurs.

Use **netstat -in** to find out the MTU of the Ethernet interface.

LAB REPORT   What is the maximum value of *size* for which the UDP datagram can be sent without IP fragmentation? Justify your answer with the **netstat** output.

| Exercise 4 | Capture the data packets generated by the following command using **tcpdump src host** *your_host*.

> **sock -u -i -n1 -w10000** *remote_host* **echo**

Save the **tcpdump** output for the lab report.

LAB REPORT   Explain the **tcpdump** output in terms of the IP header fields that are used in fragmentation.

When IP fragmentation occurs, only the first fragment has the UDP header. How do you verify this fact from the **tcpdump** output?

| Exercise 5 | While running **tcpdump src host** *your_host*, execute the following command with different values of *size*,

> **sock -u -i -n1 -w***size* *remote_host* **echo**

in order to find out the maximum size of a UDP datagram that the system can send or receive, even when fragmentation is allowed.

LAB REPORT   What is the maximum size of user data in a UDP datagram that the system can send or receive, even when fragmentation is allowed?

5.7 Path MTU discovery exercise

| Exercise 6 | In this exercise, students are divided into two groups.

Connect the routers and the workstations as shown in Fig. 5.5. Change the IP addresses of your workstation accordingly. Note that the router IP addresses are the same as their default.

Telnet to each router, enable RIP routing (see Section 4.6).

Change the MTU of the `ethernet1` interfaces of `Router2` and `Router4` to 500 bytes. To avoid confusion, each router should be configured by one person only.

Test connectivity by **ping**ing hosts in the other subnets. After you can reach the hosts in the other subnets, run **tcpdump -nx** on your workstation.

Start a UDP **sock** server on `apah` (`guchi`), using **sock -u -s 5555**.

Then run the **sock** client from `shakti` (`yachi`):

> **sock -i -u -n10 -w1200 -p5** *remote_host* **5555**,

where *remote_host* is `apah` (`guchi`).

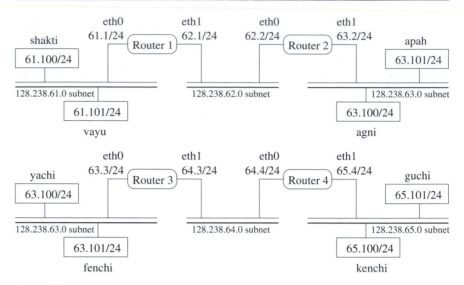

Figure 5.5. The network setup for Exercise 6.

Observe the DF bit of the first datagram and that of the following datagrams. Save the **tcpdump** output for your lab report.

Exchange **tcpdump** outputs with a student in the other subnet.

LAB REPORT   Explain the operation of path MTU discovery based on the **tcpdump** outputs saved.

Which ICMP message is used in path MTU discovery? Give the decimal value of each field of the captured ICMP message.

What is the MTU of the destination network of the UDP datagram? Verify your answer using both the ICMP message and the IP fragmentation trace saved.

5.8 Exercises with FTP and TFTP

We will study the performance of FTP and TFTP for file transfer between two machines. By transferring the same file using these two protocols, we can compare the operations and performances of UDP and TCP.

Two files (large.dum and small.dum) with random contents are stored in the /home/LAB directory and in the /tftpboot directory of each workstation in the lab. We will use the get command to retrieve files from a remote host. When FTP is used, you need to change directory to /home/LAB/ by **cd /home/LAB** before retrieving the file. If you don't know how to use **tftp**, refer to its manual page.

Exercise 7 In order to compare the transfer rates of FTP and TFTP, we will retrieve a large file from a remote server using FTP and TFTP, respectively. First run the following **tcpdump** command:

$$\text{\textbf{tcpdump host} } \textit{your\_host} \textbf{ and } \textit{remote\_host} > \textit{output1}$$

Here we use the redirect operator, >, to save the **tcpdump** output into a text file called *output1*.

Then get the /home/LAB/large.dum file from *remote_host* using **ftp**.

Also, from the **ftp** window, record the transfer rate (time) displayed.

Restart the above **tcpdump** command, with the last argument changed to output2. Now use **tftp** to get the /tftpboot/large.dum file.

Save output1 and output2 for the lab report.

LAB REPORT   Examining the saved **tcpdump** output file, output1. Identify the starting and ending time of actual data transfer. Don't include the time spent establishing the TCP connection. Calculate the time spent for data transfer.

Compare the time with the value displayed in **ftp** window. Are they consistent? If there exists any significant difference, what might be the reason?

Now, from the saved output2, carefully determine the starting and ending time of data transfer for the **tftp** program.

Compare the time with the value displayed in **tftp** window. Are they consistent? If there exists any significant difference, what might be the reason?

By comparing the actual data transfer times of **ftp** and **tftp**, which of these two is faster, and why?

Exercise 8   Capture the packets that are exchanged during a **tftp** session for the /tftpboot/small.dum file between your host and a *remote_host*, by

$$\text{\textbf{tcpdump -x host} } \textit{your\_host} \textbf{ and } \textit{remote\_host} > \textit{output3}$$

Observe the protocol in action. Analyze various types of TFTP messages used by examining the content of output3. Save output3 for the lab report.

LAB REPORT   List all the different types of packets exchanged during the **tftp** session. Compare them with the TFTP message format in Fig. 5.3.

Why does the server's port number change?

LAB REPORT   In most cases, **tftp** service is restricted.[1] Why is **tftp** service not generally available to users?

LAB REPORT   In Exercise 5, we found the maximum size of a UDP datagram in your machine. With **tftp**, which uses UDP, we transferred a file larger than the maximum UDP datagram size. How do you explain this?

Exercise 9   Repeat the above experiment, but use **ftp** and change the output file name to `output4`. Capture a trace of the packets exchanged when downloading the `/home/LAB/small.dum` file using **ftp**.

Save your **tcpdump** output. Examine the port numbers used.

LAB REPORT   How many well-known port numbers were used? Which machine used the well-known port numbers? What were the other machine's port numbers?

LAB REPORT   As can be seen from the **tcpdump** output, FTP involves two different connections, `ftp-control` and `ftp-data`. Why are two different connections used, instead of one connection?

Exercise 10   Run **ftp** in the debug mode using: **ftp -d** *remote_host*.

After logging into the *remote host*, type **dir /home/LAB/small.dum** in the **ftp** window.

Then type **quit** to terminate the **ftp** session, and save the **ftp** window output.

LAB REPORT   Submit what you saved in this exercise, explaining each line of the output.

Explain how the PORT command works.

Which connection, the control connection or the data connection, did the server send the reponse (the LIST output) on?

---

[1] This is not the case in our lab, where we deliberately enabled the TFTP service and use it as a tool to study the UDP protocol.

6   TCP study

The flow on a TCP connection should obey a 'conservation of packets' principle.
··· A new packet isn't put into the network until an old packet leaves.

Van Jacobson

6.1 Objectives

- TCP connection establishment and termination.
- TCP timers.
- TCP timeout and retransmission.
- TCP interactive data flow, using **telnet** as an example.
- TCP bulk data flow, using **sock** as a traffic generator.
- Further comparison of TCP and UDP.
- Tuning the TCP/IP kernel.
- Study TCP flow control, congestion control, and error control using DBS and NIST Net.

6.2 TCP service

TCP is the transport layer protocol in the TCP/IP protocol family that provides a *connection-oriented*, *reliable* service to applications. TCP achieves this by incorporating the following features.

- Error control: TCP uses cumulative acknowledgements to report lost segments or out of order reception, and a time out and retransmission mechanism to guarantee that application data is received reliably.
- Flow control: TCP uses sliding window flow control to prevent the receiver buffer from overflowing.

- Congestion control: TCP uses *slow start*, *congestion avoidance*, and *fast retransmit/fast recovery* to adapt to congestion in the routers and achieve high throughput.

The TCP header, shown in Fig. 0.16, consists of fields for the implementation of the above functions. Because of its complexity, TCP only supports unicast, while UDP, which is much simpler, supports both unicast and multicast. TCP is widely used in internet applications, e.g., the Web (HTTP), email (SMTP), file transfer (FTP), remote access (telnet), etc.

6.3 Managing the TCP connection

In the TCP header, the source and destination port numbers identify the sending and receiving application processes, respectively. The combination of an IP address and a port number is called a *socket*. A TCP connection is uniquely identified by the two end sockets.

6.3.1 TCP connection establishment

A TCP connection is set up and maintained during the entire session. When a TCP connection is established, two end TCP modules allocate required resouces for the connection, and negotiate the values of the parameters used, such as the *maximum segment size* (MSS), the receiving buffer size, and the *initial sequence number* (ISN). TCP connection establishment is performed by a *three-way handshake* mechanism. The TCP header format is discussed in Section 0.10.

1. An end host initiates a TCP connection by sending a packet with. ISN, n, in the sequence number field and with an empty payload field. This packet also carries the MSS and TCP receiving window size. The SYN flag bit is set in this packet to indicate a connection request.
2. After receiving the request, the other end host replies with a SYN packet acknowledging the byte whose sequence number is the ISN plus 1 ($ACK = n + 1$), and indicates its own ISN m, MSS, and TCP receiving window size.
3. The initiating host then acknowledges the byte whose sequence number is the ISN increased by 1 ($ACK = m + 1$).

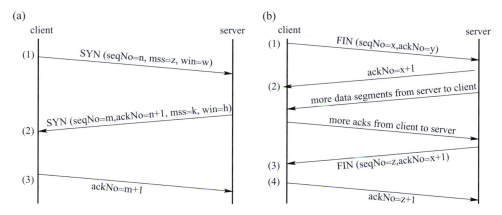

Figure 6.1. The time-line illustration of TCP connection management. (a) Three-way handshake connection establishment; (b) Four-way handshake connection termination.

After this three-way handshake, a TCP connection is set up and data transfer in both directions can begin. The TCP connection establishment process is illustrated in Fig. 6.1(a).

6.3.2 TCP connection termination

A TCP connection is full-duplex, where each end application process can transmit data to and receive data from the other end. During a TCP session, it is possible that one end application has no more data to send, while the other end does. Therefore, TCP adopts a four-way handshake to terminate the connection, giving each end of the connection a chance to shut down the one-way data flow. To do so, TCP sends a packet with the FIN flag set, and the other end acknowledges the FIN segment. This process is called the *TCP Half-Close*. After one of the data flows is shut down, the data flow in the opposite direction still works. The TCP connection is terminated only when the data flows of both directions are shut down. The TCP connection termination process is illustrated in Fig. 6.1(b).

After the final ACK [segment (4) in Fig. 6.1(b)] is sent, the connection must stay in the TIME_WAIT state for twice the maximum segment life (MSL)[1] time before termination, just to make sure that all the data on this connection has gone through. Otherwise, a delayed segment from an earlier connection may be misinterpreted as part of a new connection that uses the same local and remote sockets.

---

[1] MSL is the maximum time that any segment can exist in the network before being discarded.

If an unrecoverable error is detected, either end can close the TCP connection by sending a RST segment, where the `Reset` flag is set.

6.3.3 TCP timers

TCP uses a number of timers to manage the connection and the data flows.
- **TCP Connection Establishment Timer**. The maximum period of time TCP keeps on trying to build a connection before it gives up.
- **TCP Retransmission Timer**. If no ACK is received for a TCP segment when this timer expires, the segment will be retransmitted. We will discuss this timer in more detail in the next section.
- **Delayed ACK Timer**. Used for delayed ACK in TCP interactive data flow, which we will discuss in Section 6.4.2.
- **TCP Persist Timer**. Used in TCP flow control in the case of a fast transmitter and a slow receiver. When the advertised window size from the receiver is zero, the sender will probe the receiver for its window size when the TCP Persist Timer times out. This timer uses the normal TCP *Exponential Backoff* algorithm, but with values bounded between 5 and 60 seconds.
- **TCP Keepalive Timer**. When a TCP connection has been idle for a long time, a Keepalive timer reminds a station to check if the other end is still alive.
- **Two Maximum Segment Life Wait Timer**. Used in TCP connection termination. It is the period of time that a TCP connection keeps alive after the last ACK packet of the four-way handshake is sent [see Fig.6.1(b)]. This gives TCP a chance to retransmit the final ACK.[2] It also prevents the delayed segments of a previous TCP connection from being interpreted as segments of a new TCP connection using the same local and remote sockets.

6.4 Managing the TCP data flow

To the application layer, TCP provides a byte-stream connection. The sender TCP module receives a byte stream from the application, and puts the bytes in a sending buffer. Then, TCP extracts the bytes from the sending buffer and sends them to the lower network layer in blocks (called *TCP*

---

[2] In Fig. 6.1(b), the server will timeout if the FIN segment is not acknowledged. It then retransmits the FIN segment.

segments). The receiver TCP module uses a receiving buffer to store and re-order received TCP segments. A byte stream is restored from the receiving buffer and sent to the application process.

6.4.1 TCP error control

Since TCP uses the IP service, which is connectionless and unreliable, TCP segments may be lost or arrive at the receiver in the wrong order. TCP provides error control for application data, by retransmitting lost or errored TCP segments.

Error detection

In order to detect lost TCP segments, each data byte is assigned a unique sequence number. TCP uses *positive acknowledgements* to inform the sender of the last correctly received byte. Error detection is performed in each layer of the TCP/IP stack (by means of header checksums), and errored packets are dropped. If a TCP segment is dropped because TCP checksum detects an error, an acknowledgement will be sent to the sender for the first byte in this segment (also called the sequence number of this segment), thus effectively only acknowledging the previous bytes with smaller sequence numbers. Note that TCP does not have a negative acknowledgement feature. Furthermore, a gap in the received sequence numbers indicates a transmission loss or wrong order, and an acknowledgement for the first byte in the gap may be sent to the sender. This is illustrated in Fig. 6.2. When segment 7 is received, the receiver returns an acknowledgement for segment 8 to the sender. When segment 9 is lost, any received segment with a sequence number larger than 9 (segments 10, 11, and 12 in the example) triggers a

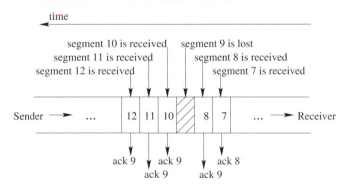

Figure 6.2. A received segment triggers the receiver to send an acknowledgement for the next segment.

duplicate acknowledgement for segment 9. When the sender receives such duplicate acknowledgements, it will retransmit the requested segment (see Section 6.4.3).

As the network link bandwidth increases, a window of TCP segments may be sent and received before an acknowledgement is received by the sender. If multiple segments in this window of segments are lost, the sender has to retransmit the lost segments at a rate of one retransmission per round trip time (RTT), resulting in a reduced throughput. To cope with this problem, TCP allows the use of *selective acknowledgement* (SACK) to report multiple lost segments. While a TCP connection is being established, the two ends can use the *TCP Sack-Permitted* option to negotiate if SACK is allowed. If both ends agree to use SACK, the receiver uses the *TCP Sack* option to acknowledge all the segments that has been successfully received in the last window of segments, and the sender can retransmit more than one lost segment at a time.

RTT measurement and the retransmission timer

On the sender side, a retransmission timer is started for each TCP segment sent. If no ACK is received when the timer expires (either the TCP packet is lost, or the ACK is lost), the segment is retransmitted.

The value of the retransmission timer is critical to TCP performance. An overly small value causes frequent timeouts and hence unnecessary retransmissions, but a value that is too large causes a large delay when a segment is lost. For best performance, the value should be larger than but of the same order of magnitude as the RTT. Considering the fact that TCP is used to connect different destinations with various RTTs, it is difficult to set a fixed value for the retransmission timer. To solve this problem, TCP continuously measures the RTT of the connection, and updates the retransmission timer value dynamically.

Each TCP connection measures the time difference between sending a segment and receiving the ACK for this segment. The measured delay is called one *RTT measurement*, denoted by M. For a TCP connection, there is at most one RTT measurement going on at any time instant. Since the measurements may have wide fluctuations due to transient congestion along the route, TCP uses a smoothed RTT, RTT^s, and the smoothed RTT mean deviation, RTT^d, to compute the retransmission timeout (RTO) value. RTT_0^s is set to the first measured RTT, M_0, while $RTT_0^d = M_0/2$ and $RTO_0 = RTT_0^s + \max\{G, 4 \times RTT_0^d\}$. G is the timeout interval of

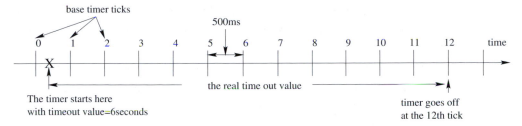

Figure 6.3. A TCP timer timeout example.

the base timer. For the ith measured RTT value M_i, RTO is updated as follows (RFC 2988):

$$RTT_i^{s} = (1 - \alpha) \times RTT_{i-1}^{s} + \alpha \times M_i, \tag{6.1}$$

$$RTT_i^{d} = (1 - \beta) \times RTT_{i-1}^{d} + \beta \times |M_i - RTT_{i-1}^{s}|, \tag{6.2}$$

$$RTO_i = RTT_i^{s} + \max\{G, 4 \times RTT_i^{d}\}, \tag{6.3}$$

where $\alpha = 1/8$ and $\beta = 1/4$. If the computed RTO is less than 1 second, then it should be rounded up to 1 second, and a maximum value limit may be placed on RTO provided that the maximum value is at least 60 seconds.

The TCP timers are discrete. In some systems, a base timer that goes off every, e.g., 500 ms, is used for RTT measurements. If there are t base timer ticks during a RTT measurement, the measured RTT is $M = t \times 500$ ms. Furthermore, all RTO timeouts occur at the base timer ticks. Figure 6.3 shows a timeout example when $RTO = 6$ seconds, and the timer goes off at the 12th base timer tick after the timer is started. Clearly the actual time out period is between 5.5 and 6 seconds. Different systems have different clock granularities. Experience has shown that finer clock granularities (e.g., $G \leq 100$ ms) perform better than more coarse granularities [8].

RTO exponential backoff

RTT measurement is not performed for a retransmitted TCP segment in order to avoid confusion, since it is not clear that if the received acknowledgement is for the original or the retransmitted segment. Both RTT^{s} and RTT^{d} are not updated in this case. This is called *Karn's Algorithm*.

What if the retransmitted packet is also lost? TCP uses the *Exponential Backoff* algorithm to update RTO when the retransmission timer expires for a retransmitted segment. The initial RTO is measured using the algorithm introduced above. Then, RTO is doubled for each retransmission, but with a maximum value of 64 seconds (see Fig. 6.4).

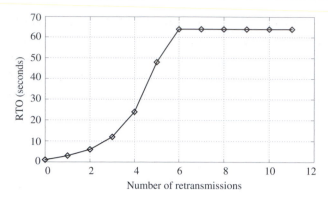

Figure 6.4. Exponential backoff of RTO after several retransmissions.

6.4.2 TCP interactive data flow

TCP supports interactive data flow, which is used by interactive user applications such as **telnet** and **ssh**. In these applications, a user keystroke is first sent from the user to the server. Then, the server echoes the key back to the user and *piggybacks* the acknowledgement for the key stroke. Finally, the client sends an acknowledgement to the server for the received echo segment, and displays the echoed key on the screen. This kind of design is effective in reducing the delay experienced by the user, since a user would prefer to see each keystroke displayed on the screen as quickly as possible, as if he or she were using a local machine.

However, a better delay performance comes at the cost of bandwidth efficiency. Consider one keystroke that generates one byte of data. The total overhead of sending one byte of application data is 64 bytes (recall that Ethernet has a minimum frame length of 64 bytes, including the TCP header, the IP header, and the Ethernet header and trailer). Furthermore, for each keystroke, three small packets are sent, resulting in a total overhead of $64 \times 3 = 192$ bytes for only 2 bytes of data (one byte from the client to the server, and one byte echoed from the server to the client). To be more efficient, TCP uses two algorithms: Delayed Acknowledgement and the Nagle algorithm, in order to reduce the number of small segments.

Delayed acknowledgement

TCP uses a delayed acknowledgement timer that goes off every K ms (e.g., 50 ms). After receiving a data segment, TCP delays sending the ACK until the next tick of the delayed acknowledgement timer, hoping that new data to be sent in the reverse direction will arrive from the application during this period. If there is new data to send during this period, the ACK can be piggybacked with the data segment. Otherwise, an ACK segment (with no data payload) is sent. Depending on when the data segment is received, when there is new data arriving from the application layer, and when the delayed acknowledgement timer goes off, an ACK may be delayed from 0 ms up to K ms.

The Nagle algorithm

The Nagle Algorithm says that each TCP connection can have only one small segment[3] outstanding, i.e., that has not been acknowledged. It can be used to further limit the number of small segments in the Internet. For interactive data flows, TCP sends one byte and buffers all subsequent bytes until an acknowledgement for the first byte is received. Then all buffered bytes are sent in a single segment. This is more efficient than sending multiple segments, each with one byte of data. But the higher bandwidth efficiency comes at the cost of increased delay for the user.

6.4.3 TCP bulk data flow

In addition to interactive flows, TCP also supports bulk data flows, where a large number of bytes are sent through the TCP connection. Applications using this type of service include email, FTP, WWW, and many others.

TCP *throughput* performance is an important issue related to the TCP bulk data flows. Ideally, a source may wish to always use the maximum sending rate, in order to deliver the application bulk data as quickly as possible. However, as discussed in Section 0.8, if there is congestion at an intermediate router or at the receiving node, the more packets a source sends, the more packets would be dropped. Furthermore, the congestion will persist until some or all of the data flows reduce their transmission rates. Therefore, for a high throughput, the source should always try to increase its sending rate. On the other hand, for a low packet loss rate, the source rate should be bounded by the maximum rate that can be allowed without causing congestion or receiver buffer overflow, and should be adaptive to network conditions.

TCP sliding window flow control

TCP uses sliding window flow control to avoid receiver buffer overflow, where the receiver advertises the maximum amount of data it can receive

---

[3] which is less than one MSS.

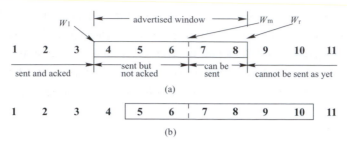

Figure 6.5. A TCP sliding window flow control example. (a) The sliding window maintained by the sender. (b) The updated sliding window when an acknowledgement, [*ackno* = 5, *awnd* = 6] is received.

(called the *Advertised Window*, or *awnd*), and the sender is not allowed to send more data than the advertised window.

Figure 6.5(a) illustrates the sliding window flow control algorithm. The application data is a stream of bytes, where each byte has a unique sequence number. In Fig. 6.5, each block represents a TCP segment with MSS bytes, and the number can be regarded as the sequence number of the TCP segments in units of MSS bytes. In TCP, the receiver notifies the sender (1) the next segment it expects to receive and (2) the amount of data it can receive without causing a buffer overflow (denoted as [*ackno* = *x*, *awnd* = *y*]), using the `Acknowledgement Number` and the `Window Size` fields in the TCP header. Figure 6.5(a) is the sliding window maintained at the sender. In this example, segments 1 through 3 have been sent and acknowledged. Since the advertised window is five segments and the sender already has three outstanding segments (segments 4, 5, and 6), at most two more segments can be sent before a new acknowledgement is received.

The sliding window, shown as a box in Fig. 6.5, moves to the right as new segments are sent, or new acknowledgements and window advertisements are received. More specifically, if a new segment is acknowledged, W_l, the left edge of the window, will move to the right (*window closes*). W_m moves to the right when new segments are sent. If a larger window is advertised by the receiver or when new segments are acknowledged, the right edge of the sliding window, W_r, will move to the right (*window opens*). However, if a smaller window is advertised, W_r will move to the left (*window shrinks*). Figure 6.5(b) illustrates the updated sliding window when an acknowledgement, [*ackno* = 5, *awnd* = 6], is received.

With this technique, the sender rate is effectively determined by (1) the advertised window, and (2) how quickly a segment is acknowledged. Thus a slow receiver can advertise a small window or delay the sending of acknowledgements to slow down a fast sender, in order to keep the receiver buffer from overflowing. However, even with effective flow control, a TCP segment may still be dropped at an intermediate router when the router buffer is full due to congestion. In addition to sliding window flow control, TCP uses congestion control to cope with network congestion.

TCP congestion control

TCP uses congestion control to adapt to network congestion and achieve a high throughput. Usually the buffer in a router is shared by many TCP connections and other non-TCP data flows, since a shared buffer leads to a more efficient buffer utilization and is easier to implement than assigning a separate buffer for each flow. TCP needs to adjust its sending rate in reaction to the rate fluctuations of other data flows sharing the same router buffer. In other words, a new TCP connection should increase its rate as quickly as possible to take all the available bandwidth. When the sending rate is higher than some threshold, TCP should slow down its rate increase to avoid congestion.

Considering the huge number of TCP connections going through an Internet core router, routers are designed to be as simple as possible. Usually a router simply drops incoming packets when its buffer is full, without notifying the sender. However, the sender can infer congestion along the route when a retransmission timer goes off. In addition, the receiver also reports congestion to the sender implicitly by sending duplicate acknowledgements (see Fig. 6.2). When congestion occurs, TCP drastically reduces its sending rate. The reason is that if the router is congested, the more data sent, the more data would be dropped. Furthermore, if some of the TCP connections lose packets and reduce their rates, it is likely that the congestion will abate and disappear.

More specifically, the sender maintains two variables for congestion control: a congestion window size ($cwnd$), which upper bounds the sender rate, and a slow start threshold ($ssthresh$), which determines how the sender rate is increased. The TCP *slow start* and *congestion avoidance* algorithms are given in Table 6.1. According to these algorithms, $cwnd$ initially increases exponentially until it reaches $ssthresh$. After that, $cwnd$ increases roughly linearly. When congestion occurs, $cwnd$ is reduced to 1 MSS to avoid segment loss and to alleviate congestion. It has been shown that when N TCP connections with similar RTTs share a bottleneck router with an output link bandwidth of C, their long-term average rates quickly converge to the optimal operating rates, i.e., each TCP connection has an

Table 6.1. *The slow start and congestion avoidance algorithms*

(1)   If $cwnd \leq ssthresh$ then   /* *Slow Start* Phase */

     Each time an ACK is received:

       $cwnd = cwnd + segsize$

     else /* *Congestion Avoidance* Phase */

     Each time an ACK is received:

       $cwnd = cwnd + segsize \times segsize/cwnd + segsize/8$

     end

(2)   When congestion occurs (indicated by retransmission timeout)

       $ssthresh = \max(2, \min(cwnd, awnd)/2)$

       $cwnd = 1 \; segsize = 1 \; MSS \; bytes$

(3) $Allowed \; window = \min(cwnd, awnd)$

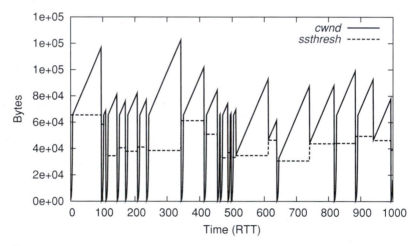

Figure 6.6. The evolution of $cwnd$ and $ssthresh$ for a TCP connection, including slow start, congestion avoidance, fast retransmit, and fast recovery.

average rate of C/N, when this *additive-increase-multiplicative-decrease* (AIMD) algorithm is used [9]. Another advantage of this algorithm is that it is *self-clocking*. The higher the rate at which acknowledgements are received (which implies that the congestion is light), the quicker the sending rate increases. Figure 6.6 illustrates the evolution of $cwnd$ and $ssthresh$ of a TCP connection. It can be seen clearly that the evolution of $cwnd$ has two phases, i.e., an exponential increase phase and a linear increase phase. When there is a packet loss, $cwnd$ drops drastically.

TCP allows accelerated retransmissions. Recall that when there is a gap in the receiving buffer, the receiver will acknowledge the first byte in the

Table 6.2. *TCP fast retransmit/fast recovery algorithm*

(1)	After the third duplicate ACK is received:
	$ssthresh = cwnd/2$
	retransmit the missing segment
	$cwnd = ssthresh + 3segsize$
(2)	For each additional duplicate acknowledgement received:
	$cwnd = cwnd + segsize$
	transmit a segment if allowed by $cwnd$
(3)	When the acknowledgement for the retransmitted segment arrives:
	$cwnd = ssthresh + segsize$

gap. Further arriving segments, other than the segment corresponding to the gap, trigger duplicate acknowledgements (see Figure 6.2). After receiving three duplicate acknowledgements, the sender assumes that the segment is lost and retransmit the segment immediately without waiting for the retransmission timer to expire. This algorithm is called the *fast retransmit* algorithm. After the retransmission, congestion avoidance, rather than slow start, is performed, with an initial *cwnd* equal to *ssthresh* plus one segment size.[4] This is called the *fast recovery* algorithm. With these two algorithms, *cwnd* and *ssthresh* are updated as shown in Table 6.2. In the example shown in Fig. 6.6, TCP fast retransmit and fast recovery occur at time instances around 610, 740, and 950.

6.5 Tuning the TCP/IP kernel

TCP/IP uses a number of parameters in its operations (e.g., TCP keepalive timer). Since the TCP/IP protocols are used in many applications, a set of default values may not be optimal for different situations. In addition, the network administrator may wish to turn on (or off) some TCP/IP functions (e.g., ICMP redirect) for performance or security considerations. Many Unix and Linux systems provide some flexibity in tuning the TCP/IP kernel.

In Red Hat Linux, /sbin/sysctl is used to configure the Linux kernel parameters at runtime. The default kernel configuration file is /sbin/sysctl.conf, consisting of a list of kernel parameters and their

---

[4] The duplicate acknowledgements imply that the subsequent segments have been received. Therefore, the network is not congested and the sender need not switch to the slow start phase to reduce its rate.

default values. For the parameters with binary values, a "0" means the function is disabled, while a "1" means the function is enabled. Some frequently used **sysctl** options are listed here.

- **sysctl -a** or **sysctl -A**: list all current values.
- **sysctl -p** *file_name*: to load the sysctl setting from a configuration file. If no file name is given, `/etc/sysctl.conf` is used.
- **sysctl -w** *variable=value*: change the value of the parameter.

The TCP/IP related kernel parameters are stored in the `/proc/sys/net/ipv4/` directory. As an alternative to the **sysctl** command, you can modify these files directly to change the TCP/IP kernel setting. For example, the default value of the TCP keepalive timer is saved in the `/proc/sys/net/ipv4/tcp_keepalive_time` file. As root, you can run

echo '3600' > /proc/sys/net/ipv4/tcp_keepalive_time

to change the TCP keepalive timer value from its default 7200 seconds to 3600 seconds.

Solaris 8.0 provides a program, **ndd**, for tuning the TCP/IP kernel, including the IP, ICMP, ARP, UDP and TCP modules. To display a list of parameters editable in a module, use the following command:

ndd *module* \?,

where *module* could be `/dev/ip`, `/dev/icmp`, `/dev/arp`, `/dev/udp`, or `/dev/tcp`. To display the current value of a parameter, use:

ndd -get *module parameter*.

To modify the value of a parameter in a module, use:

ndd -set *module parameter*.

6.6 TCP diagnostic tools

6.6.1 The distributed benchmark system

The distributed benchmark system (DBS) is a benchmark for TCP performance evaluation. It can be used to run tests with multiple TCP connections or UDP flows and to plot the test results. DBS consists of three tools.

- **dbsc**: the DBS test controller.

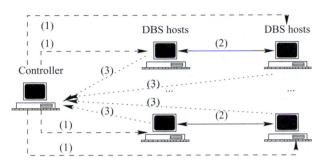

Figure 6.7. The operation of DBS.

- **dbsd**: the DBS daemon, running on each participating host.
- **dbs_view**: a `Perl` script file, used to plot the experiment results.

    DBS uses a command file to describe the test setting. In the command file, a user can specify (1) how many TCP or UDP flows to generate, (2) the sender and receiver for each flow, (3) the traffic pattern and duration of each flow, and (4) which statistics to collect. During a test, one host serves as the controller, running **dbsc**, and all other participating hosts are DBS hosts, running **dbsd**. As illustrated in Fig. 6.7, the controller first reads the command file and sends instructions to all the DBS hosts. Second, TCP (or UDP) connections will be set up between the DBS hosts and TCP (or UDP) traffic is transmitted on these connections as specified in the command file. Third, when the data transmissions are over, the DBS controller collects statistics from the DBS hosts which may be plotted using **dbs_view**.

6.6.2 NIST Net

NIST Net is a Linux-based network emulator. It can be used to emulate various network conditions, such as packet loss, duplication, delay and jitter, bandwidth limitations, and network congestion. As illustrated in Fig. 6.8, a Linux host running NIST Net serves as a router between two subnets. There are a number of TCP connections or UDP flows traversing this router host. NIST Net works like a firewall. A user can specify a connection, by indicating its source IP and destination IP, and enforce a policy, such as a certain delay distribution, a loss distribution, or introduce packet duplication on this connection.

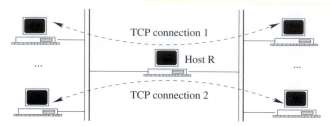

Figure 6.8. The operation of NIST Net.

6.6.3 Tcpdump output of TCP packets

Generally, **tcpdump** outputs a captured TCP packet in the following format.

> timestamp   src_IP.src_port > dest_IP.dest_port: flags seq_no ack window urgent options

The following is a sample **tcpdump** output, which shows a TCP packet captured at time 54:16.401963 (Minute:Second:MicroSecond). The TCP connection is between aida.poly.edu and mng.poly.edu, with source TCP port 1121 and destination TCP port telnet (23). The PUSH flag bit is set. The sequence number of the first data byte is 1,031,880,194, and 24 bytes of data is carried in this TCP segment. aida is expecting byte 172,488,587 from mng and advertises a window size of 17,520 bytes.

> 54:16.401963   aida.poly.edu.1121 > mng.poly.edu.telnet: P 1031880194
> :1031880218(24) ack 172488587 win 17520

6.7 Exercises on TCP connection control

Exercise 1 While **tcpdump -S host** *your_host* **and** *remote_host* is running, execute: **telnet** *remote_host* **time**.

Save the **tcpdump** output.

LAB REPORT   Explain TCP connection establishment and termination using the **tcpdump** output.

LAB REPORT   What were the announced MSS values for the two hosts?

What happens if there is an intermediate network that has an MTU less than the MSS of each host?

See if the DF flag was set in our **tcpdump** output.

Exercise 2 While **tcpdump -nx host** *your_host* **and** *remote_host* is running, use **sock** to send a UDP datagram to the remote host:

> **sock -u -i -n1** *remote_host* **8888**.

Save the **tcpdump** output for your lab report.

Restart the above **tcpdump** command, execute **sock** in the TCP mode:

> **sock -i -n1** *remote_host* **8888**.

Save the **tcpdump** output for your lab report.

LAB REPORT   Explain what happened in both the UDP and TCP cases. When a client requests a nonexisting server, how do UDP and TCP handle this request, respectively?

6.8 Exercise on TCP interactive data flow

Exercise 3 While **tcpdump** is capturing the traffic between your machine and a remote machine, issue the command: **telnet** *remote_host*.

After logging in to the host, type date and press the Enter key.

Now, in order to generate data faster than the round-trip time of a single byte to be sent and echoed, type any sequence of keys in the **telnet** window very rapidly.

Save the **tcpdump** output for your lab report. To avoid getting unwanted lines from **tcpdump**, you and the student who is using the remote machine should do this experiment in turn.

LAB REPORT   Answer the following questions, based upon the **tcpdump** output saved in the above exercise.

(1) What is a delayed acknowledgement? What is it used for?

(2) Can you see any delayed acknowledgements in your **tcpdump** output?

If yes, explain the reason. Mark some of the lines with delayed acknowledgements, and submit the **tcpdump** output with your report.

Explain how the delayed ACK timer operates from your **tcpdump** output.

If you don't see any delayed acknowledgements, explain the reason why none was observed.

(3) What is the Nagle algorithm used for?

From your **tcpdump** output, can you tell whether the Nagle algorithm is enabled or not? Give the reason for your answer.

From your **tcpdump** output for when you typed very rapidly, can you see any segment that contains more than one character going from your workstation to the remote machine?

6.9 Exercise on TCP bulk data flow

Exercise 4   While **tcpdump** is running and capturing the packets between your machine and a remote machine, on the remote machine, which acts as the server, execute: **sock -i -s 7777**.

Then, on your machine, which acts as the client, execute:

sock -i -n16 *remote_host* **7777**.

Do the same experiment three times. Save all the **tcpdump** outputs for your lab report.

LAB REPORT   Using one of three **tcpdump** outputs, explain the operation of TCP in terms of data segments and their acknowledgements. Does the number of data segments differ from that of their acknowledgements?

Compare all the **tcpdump** outputs you saved. Discuss any differences among them, in terms of data segments and their acknowledgements.

LAB REPORT   From the **tcpdump** output, how many different TCP flags can you see? Enumerate the flags and explain their meanings.

How many different TCP options can you see? Explain their meanings.

6.10 Exercises on TCP timers and retransmission

Exercise 5   Execute **sysctl -A | grep keepalive** to display the default values of the TCP kernel parameters that are related to the TCP keepalive timer.

What is the default value of the TCP keepalive timer? What is the maximum number of TCP keepalive probes a host can send?

In Solaris, execute **ndd -get /dev/tcp tcp_keepalive_interval** to display the default value of the TCP keepalive timer.

LAB REPORT   Answer the above questions.

Exercise 6 | While **tcpdump** is running to capture the packets between your host and a remote host, start a **sock** server on the remote host, **sock -s 8888**.

Then, execute the following command on your host,

sock -i -n200 *remote_host* **8888**.

While the sender is injecting data segments into the network, disconnect the cable connecting the sender to the hub for about ten seconds.

After observing several retransmissions, reconnect the cable. When all the data segments are sent, save the **tcpdump** output for the lab report.

LAB REPORT   Submit the **tcpdump** output saved in this exercise.

From the **tcpdump** output, identify when the cable was disconnected.

Describe how the retransmission timer changes after sending each retransmitted packet, during the period when the cable was disconnected.

Explain how the number of data segments that the sender transmits at once (before getting an ACK) changes after the connection is reestablished.

6.11 Other exercises

Exercise 7 | While **tcpdump src host** *your_host* is running, execute the following command, which is similar to the command we used to find out the maximum size of a UDP datagram in Chapter 5,

sock -i -n1 -w*n* *host* **echo**

Let *n* be larger than the maximum UDP datagram size we found in Exercise 5 of Chapter 5. As an example, you may use *n* = 70,080.

LAB REPORT   Did you observe any IP fragmentation?

If IP fragmentation did not occur this time, how do you explain this compared to what you observed in Exercise 5 of Chapter 5?

Exercise 8 | Study the manual page of **/sbin/sysctl**. Examine the default values of some TCP/IP configuration parameters that you might be interested in. Examing the configuration files in the /proc/sys/net/ipv4 directory.

When Solaris is used, use **ndd** to examine the TCP/IP configuration parameters. See Section 6.5 or the manual page of **ndd** for the syntax and parameters.

Table 6.3. *Two groups for exercises in Section 6.12*

-	host1	host2	host3	host4
Group A	shakti	vayu	agni	apah
Group B	yachi	fenchi	guchi	kenchi

6.12 Exercises with DBS and NIST Net

In this exercise, students are divided into two groups as shown in Table 6.3. The four hosts in each group are connected by a hub. All the hosts have the default IP addresses and subnet masks as shown in Table 1.2.

Before these exercises, the lab instructor should start **ntpd** to synchronize the hosts. First, modify the /etc/ntp.conf file in all the hosts as follows: (1) comment the "restrict default ignore" line, and (2) for host1, host2, and host3 in Group A, insert a new line "server 128.238.66.103"; for host1, host2, and host3 in Group B, insert a new line "server 128.238.66.107". For example, the /etc/ntp.conf file in host1, host2, and host3 look should like the following:

```
. . .
# restrict default ignore
. . .
server 128.238.66.103 # for Group A
# server 128.238.66.107 # for Group B
. . .
```

Second, start the **ntpd** daemon by running **/etc/init.d/ntpd start**. Then all the hosts in Group A (Group B) will be synchronized with apah (kenchi). Note that it may take a while (several minutes) for the hosts to be synchronized, since by default an NTP client polls a server every 60 seconds.

Exercise 9 In the following, we will use DBS to study the performance of TCP under different background traffic. The DBS command files used in this exercise are given in Appendix C.1.

The TCP1.cmd file in Section C.1.1 of Appendix C1 is used to set up a TCP connection between host1 and host2, where host2 sends a stream of packets to host1. Edit the TCP1.cmd file, replace the values for the hostname variable to the IP addresses of the corresponding hosts in your group as shown in Table 6.3. For example, in group A, host1 is `shakti` and host2 is `vayu`. So the TCP1.cmd for Group A should be changed as shown below:

```
...
sender {
        hostname = 128.238.66.101
...
receiver {
        hostname = 128.238.66.100
...
```

In all the following experiments, we will use `host4` as the DBS controller. Start **tcpdump host** *host1_IP* **and** *host2_IP* on all the hosts. Then start **dbsd** on all other hosts except host4 (`apah` in Group A and `kenchi` in Group B). Next, execute **dbsc TCP1.cmd** on host4.

Observe the data transmissions between `host1` and `host2` from the **tcpdump** output.

When the data transmission is over, execute the following two commands on host4 to plot the received sequence numbers and throughput of the TCP connection:

> **dbs_view -f TCP1.cmd -sq sr -p -ps -color** > **ex9sqa.ps**,
> **dbs_view -f TCP1.cmd -th r -p -ps -color** > **ex9tha.ps**.

Save these two Postscript files for the lab report. You can use the GIMP graphical tool in Red Hat Linux to convert the Postscript files to other formats. The second **dbs_view** command also gives the average throughput of the TCP connection. Save this number for the lab report.

Next, edit the `TCPUDP.cmd` file given in Section C.1.2 of Appendix C. Replace the `hostname` fields with the corresponding IP addresses for the senders and the receivers according to Table 6.3. Then, repeat the above exercise, but use the `TCPUDP.cmd` file. This file consists of commands to start a TCP connection with the same parameters as the previous exercise, and a UDP flow emulating an MPEG video download. Oberve the impact on TCP performance of the UDP flow.

When the data transmission is over, execute the following two commands to plot the received sequence numbers and throughput of the TCP connection:

> **dbs_view -f TCPUDP.cmd -sq sr -p -ps -color** > **ex9sqb.ps**,
> **dbs_view -f TCPUDP.cmd -th r -p -ps -color** > **ex9thb.ps**.

Save these two Postscript files, as well as the average throughputs of the TCP connection and the UDP flow.

Table 6.4. *The NIST Net settings for Exercise 10*

Source	Dest	Delay (ms)
host1_IP	host2_IP	20 ms
host3_IP	host2_IP	500 ms

LAB REPORT   Compare the throughput of the TCP connections in the above two experiments. In which case does the TCP connection have higher throughput?
Justify you answer with the throughput plots and the sequence number
plots.

> **Exercise 10**   [5]In one command window, execute **tcpdump ip host** *host1_IP* **and** *host2_IP* to
> capture the TCP packets between host1 and host2. In another command window,
> run **tcpdump ip host** *host3_IP* **and** *host2_IP* to capture the TCP packets between
> host3 and host2.
>
> On host1, execute **Load.Nistnet** to load the NIST Net emulator module into the Linux
> kernel.
>
> Execute **xnistnet** on host1 (shakti in Group A and yachi in Group B). Enter the
> values in the NIST Net GUI as given in Table 6.4. Then click the Update button
> to enforce a 20 ms delay on the TCP connection between host1 and host2, and a
> 500 ms delay on the TCP connection between host2 and host3.
>
> Start the DBS daemon on host1, host2, and host3, by running **dbsd -d**.
>
> Edit the TCP2.cmd file given in Section C.1.3 of Appendix C on host4. Set the
> hostname values in the command file to the corresponding IP addresses according
> to Table 6.3. Execute the DBS controller on host4, by **dbsc TCP2.cmd**.
>
> Observe the data transmissions shown in the **tcpdump** outputs. When data trans-
> missions are over, save the **tcpdump** outputs and use the following command to
> plot the received sequence numbers and throughputs of the two TCP connections:
>
> > **dbs_view -f TCP2.cmd -sq sr -p -ps -color** > **ex10sq.ps**,
> > **dbs_view -f TCP2.cmd -th r -p -ps -color** > **ex10th.ps**,
>
> Save the plots and the mean throughputs of the two TCP connections from the
> **dbs_view** outputs.

LAB REPORT   From the received sequence number plot, can you tell which TCP
connection has higher throughput? Why? Justify your answer using the
tcpdump outputs and the **dbs_view** plots.

---

[5] This exercise is for Linux only, since NIST Net does not run on Solaris.

Exercise 11 | [6]Restart the **xnistnet** program on `host1`. Set `Source` to `host2`'s IP address and `Dest` to `host1`'s IP address. Set `Delay` for this connection to be 500 ms, and `Delsigma` to 300 ms. This enforces a mean delay of 500 ms and a delay deviation of 300 ms for the IP datagrams between `host1` and `host2`.

Execute **tcpdump ip host** *host1_IP* **and** *host2_IP* on all the hosts.

Start a **sock** server on `host1` by running **sock -i -s 7777**. Start a **sock** client on `host2` by running **sock -i -n50** *host1_IP* **7777** to pump TCP packets to `host1`.

When the data transfer is over, examine the **tcpdump** outputs to see if a retransmission or fast retransmission occured. If you cannot see one, you may try running the **sock** program again.

LAB REPORT   Submit the section of a **tcpdump** output saved that has out of order TCP segments arriving at the receiver.

Exercise 12 | [7]This exercise is similar to the previous one, except that `Delay` is set to 100 ms, `Delsigma` is set to 0 ms, and `Drop` is set to 5%.

Run the **sock** server and client. When the data transfer is over, examine the **tcpdump** output. Can you see any packet loss and retransmission? Justify your answer using the **tcpdump** output.

Try different values for the `Drop` field, or different combinations of `Delay`, `DelSigma`, and `Drop`.

LAB REPORT   Answer the above questions.

---

[6] This exercise is for Linux only, since NIST Net does not support Solaris.
[7] This exercise is for Linux only, since NIST Net does not support Solaris.

7 Multicast and realtime service

We are now in a transition phase, just a few years shy of when IP will be the universal platform for multimedia services.

H. Schulzrinne

7.1 Objectives

- Multicast addressing.
- Multicast group management.
- Multicast routing: configuring a multicast router.
- Realtime video streaming using the Java Media Framework.
- Protocols supporting realtime streaming: RTP/RTCP and RTSP.
- Analyzing captured RTP/RTCP packets using Ethereal.

7.2 IP multicast

IP provides three types of services, i.e., *unicast*, *multicast*, and *broadcast*. Unicast is a *point-to-point* type of service with one sender and one receiver. Multicast is a *one-to-many* or *many-to-many* type of service, which delivers packets to multiple receivers. Consider a multicast group consisting of a number of participants, any packet sent to the group will be received by all of the participants. In broadcasts, IP datagrams are sent to a broadcast IP address, and are received by all of the hosts.

Figure 7.1 illustrates the differences between multicast and unicast. As shown in Fig. 7.1(a), if a node A wants to send a packet to nodes B, C, and D using unicast service, it sends three copies of the same packet, each with a different destination IP address. Then, each copy of the packet will follow a possibly different path from the other copies. To provide a teleconferencing-type service for a group of N nodes, there need to be

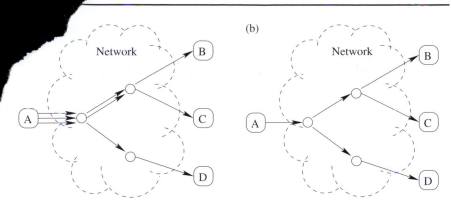

(b)

Figure 7.1. Comparison of IP unicast and multicast. (a) A unicast example, where node A sends three copies of the same packet to nodes B, C, and D. (b) A multicast example, where node A sends a packet to the multicast group, which consists of nodes B, C, and D.

$N(N-1)/2$ point-to-point paths to provide a full connection. On the other hand, if multicast service is used, as illustrated in Fig. 7.1(b), node A only needs to send one copy of the packet to a common *group address*.[1] This packet will be forwarded or replicated in a multicast tree where node A is the root and nodes B, C, D are the leaves. All nodes in this group, including node B, C, and D, will receive this packet. With multicast, clearly less network resources are used.

IP multicast is useful in providing many network services, e.g., naming (DNS), routing (RIP-2), and network management (SNMP). In many cases, it is used when a specific destination IP address is unknown. For example, in the ICMP router discovery exercise in Chapter 4, a host sends an ICMP router solicitation message to a multicast group address meaning `all routers in this subnet`. All routers connecting to this subnet receive this request, although the host may not know if there are any routers out there, and if there are, what IP addresses their interfaces have. In addition, IP multicast is widely used in multimedia streaming (e.g., video conferencing and interactive games) due to its efficiency. As illustrated in Fig. 7.1, a multicast group (consisting of nodes A, B, C, D) is easier to manage and uses less network resources than providing an end-to-end connection between every two participating nodes.

The example in Fig. 7.1(b) illustrates the three key components in providing multicast services.

---

[1] RFC 1112 indicates that the sender, e.g. node A, does not have to be in the multicast group.

1. Multicast addressing. How to define a common group nodes in the group to use, and how to map a multicast group a MAC address.

2. Multicast group management. The multicast group is dynamic, that users may join and leave the group during the multicast s A multicasting router needs to keep track of the memberships multicast groups, and a participant may want to know who else is i group.

3. Multicast routing. A multicast tree should be found and maintained from a participating node to all other nodes in the group, and the tree should be updated when either the network topology changes or the group membership changes.

We will examine these three key components of IP multicasting in the following sections.

7.2.1 Multicast addressing

IP multicast addressing

One salient feature of IP multicast is the use of a group IP address instead of a simple destination IP address. A multicast group consists of a number of participating hosts and is identified by the group address. A multicast group can be of any size, and the participants can be at various geographical locations.

In the IP address space, Class D addresses are used for multicast group addresses, ranging from 224.0.0.0 to 239.255.255.255. There is no structure within the Class D address space. This is also different from unicast IP addresses, where the address field is divided into three sub-fields, i.e., network ID, subnet ID, and host ID. However, some segments of Class D addresses are well-known or reserved. For example, all the Class D addresses between 224.0.0.0 and 224.0.0.255 are used for local network control, and all the Class D addresses between 224.0.1.0 and 224.0.1.255 are used for internetwork control. Table 7.1 gives several examples of the well-known Class D addresses. For example, in Exercise 5 of Chapter 4, a host sends an ICMP router discovery request to the Class D address 224.0.0.2, which is the group ID of all the router interfaces in a subnet.

Ethernet multicast addressing

A 48-bit long Ethernet address consists of a 23-bit *vendor component*, a 24-bit *group identifier* assigned by the vendor, and a *multicast bit*, as illustrated

Table 7.1. *Examples of reserved multicast group addresses*

224.0.0.1	All systems in this subnet
224.0.0.2	All routers in this subnet
224.0.0.4	All Distance Vector Multicast Routing ... routers in this subnet
224.0.0.5	All Multicast extension to OSPF routers ...et
224.0.0.9	Used for RIP-2
224.0.0.13	All Protocol Independent Multicast route...
224.0.1.1	Used for the Network Time Protocol

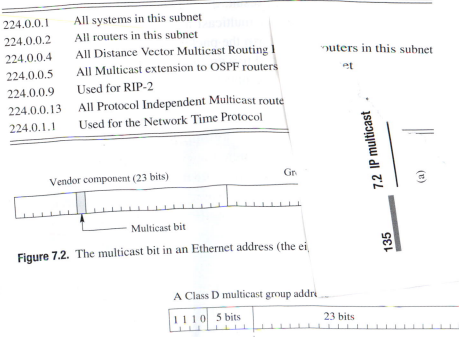

Vendor component (23 bits)

Gr...

Multicast bit

Figure 7.2. The multicast bit in an Ethernet address (the ei...

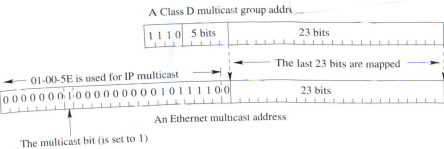

A Class D multicast group addr...

| 1 1 1 0 | 5 bits | 23 bits |

The last 23 bits are mapped

01-00-5E is used for IP multicast

0 0 0 0 0 0 0 1 0 0 0 0 0 0 0 0 0 0 1 0 1 1 1 1 1 0 0 | 23 bits

An Ethernet multicast address

The multicast bit (is set to 1)

Figure 7.3. Mapping a Class D multicast IP address to an Ethernet multicast address.

in Fig. 7.2. The vendor block is a block of Ethernet addresses assigned to a vendor. For example, Cisco is assigned with the vendor component 0x00-00-0C. Thus all the Ethernet cards made by Cisco have Ethernet addresses starting with this block. The multicast bit is used to indicate if the current frame is multicast or unicast. If the multicast bit is set, this Ethernet address is a multicast Ethernet address. Therefore, a multicast Ethernet address assigned to Cisco starts with 0x01-00-0C.

Multicast address mapping

The Ethernet address segment starting with 0x01-00-5e is used for IP multicasting. When there is a multicast packet to send, the multicast destination IP address is directly mapped to an Ethernet multicast address. No ARP request and reply are needed. The mapping is illustrated in Fig. 7.3. Note that

only the last 23 bits of
MAC address. As a re
to the same Ethernet
module should perfor
IP datagrams destine

At the receiver, tl
module to join or le
of group membersl
joins a new group or leaves a gro
be able to join or leave a multicast group. When the net
new group, its *reception filters* are modified to enable reception of multicast
Ethernet frames belonging to the group. A router interface should then be
able to receive all the multicast IP datagrams.

7.2.2 Multicast group management

The Internet Group Management Protocol (IGMP) is used to keep track
of multicast group memberships in the last hop of the multicast tree. A
host uses IGMP to announce its multicast memberships, and a router uses
IGMP to query multicast memberships in the attached networks. Figure 7.4
shows the IGMP version 1 message format. An IGMP message is eight
bytes long. The Type field is set to 1 for a query sent by a router, and 2 for
a report sent by a host. The last four bytes carry a multicast group address.
For the IGMPv2 message format in Fig. 7.5, the possible Type values
are: 0x11 for membership query, 0x16 for version 2 membership report,
0x17 for leaving the group, and 0x12 for version 1 membership report
to maintain backward-compatibility with IGMPv1. The Max Resp Time,

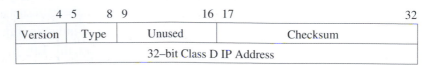

Figure 7.4. The IGMP version 1 message format.

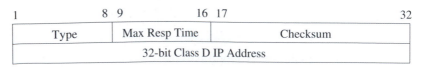

Figure 7.5. The IGMP version 2 message format.

DVMRP uses a *flood-*
grams. In DVMRP, a sour
the network. A DVMRP
S if, and only if the follow
• The packet comes from
  is called *Reverse Path F*
• *R* forwards the packet on
  is defined as the link that
  S is the root. The child li
  Thus, a multicast datagram
  the shortest path tree with *S*
  values to the TTL
  broadca

which is applicable only to query messages, specifies the maximum allowed time before sending a report message, in units of 1/10 seconds.

With IGMP, multicast routers periodically send host membership queries to discover which multicast groups have members on their attached local networks. By default, the queries are transmitted at 60 second intervals. These queries are sent to the Class D address 224.0.0.1 (*all hosts in the subnet*) with a TTL of 1. When a host receives an IGMP query, it responds with an IGMP report for each multicast group in which it is a member. The destination IP address of the IP datagram carrying the IGMP report is identical to the multicast group it is reporting on. Recall that a router interface receives all multicast datagrams. In order to avoid a flood of reports, a host delays an IGMP report for a random amount of time. During this interval, if it overhears a report reporting on the same group address, it cancels the transmission of the report. Thus the total number of reports transmitted is suppressed. When a host leaves a multicast group, it may do so silently and its membership record at the router will expire and be removed. Later versions of IGMP (e.g., IGMPv2 or IGMPv3) allow a host to report to all the routers when it leaves a multicast group (type value is 0x17).

A multicast router maintains a multicast group membership table. The table records which groups have members in the local networks attached to each of the router interfaces. The router uses the table to decide which ports to forward a multicast datagram to.

7.2.3 Multicast routing

In IP multicast, participants in a group could be in different geographical locations. A user can join and leave the multicast session at will. The size of a group could be from 1 to an arbitrarily large number. These make multicast routing more difficult than unicast routing. A critical issue in supporting IP multicast is how to find the trees for distributing multicast IP datagrams at a moderate cost (in terms of both network bandwidth resources and router CPU and memory usage).

Distance Vector Multicast Routing Protocol (DVMRP)

As suggested by its name, DVMRP is a distance vector-based multicast routing protocol. A DVMRP router exchanges multicast routing information with its neighbors, and builds the multicast routing table based on these multicast routing updates.

DVMRP uses a *flood-and-prune* approach in routing multicast IP datagrams. In DVMRP, a source broadcasts the first multicast IP datagram over the network. A DVMRP router R forwards a multicast packet from source S if, and only if the following conditions apply.

- The packet comes from the shortest route from R back to S. This scheme is called *Reverse Path Forwarding*.
- R forwards the packet only to the *child links* for S. A child link of R for S is defined as the link that has R as parent on the shortest path tree where S is the root. The child links are found by multicast routing updates.

Thus, a multicast datagram is effectively flooded to the entire network using the shortest path tree with S as the root. In addition, DVMRP assigns various values to the TTL field of multicast datagrams to control the scope of the broadcast. Furthermore, each link can be assigned with a *TTL threshold* in addition to the routing cost. A router will not forward a multicast/broadcast datagram if its TTL is less than the threshold.

When the packet arrives at a router with no record of membership in that group, the router will send a *prune* message, or a *non-membership report*, upstream of the tree, so that the branch will be deleted from the multicast tree. On the other hand, when a new member in a pruned subnet joins the group, the new membership will be detected by the router using IGMP. Next, the router will send a message to the core to undo the prune. This technique is called *grafting*.

As in RIP, DVMRP is based on the distance vector routing algorithm. Therefore, it has the same limitations as RIP, e.g., it also has the count-to-infinity problem. DVMRP uses multiple multicast trees, each with a source as its root. The multicast routing daemon for DVRMP is **mrouted**.

Multicast extension to OSPF (MOSPF)

MOSPF is an *intra-domain* multicast routing protocol, i.e., it finds multicast trees within an AS. Recall that as described in Section 4.2.4, OSPF uses LSAs to exchange link state information. In MOSPF, a new LSA called *group membership LSA* is used. In addition to other types of LSAs, multicast routers also flood group membership LSAs to distribute group membership information on the attached networks. A MOSPF router then computes the shortest path tree to all other subnets with at least one member of the multicast group.

As in DVMRP, MOSPF uses multiple multicast trees, each with a source as the root. In order to reduce the routing overhead, both DVMRP and

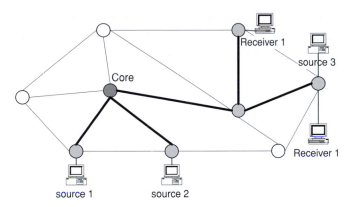

Figure 7.6. A shared multicast tree.

MOSPF perform the tree computation *on-demand*, i.e., the computation is triggered by the first arriving multicast datagram to a group.

Core-based tree (CBT)

Both DVMRP and MOSPF use one multicast tree for each source. This could be very costly when the network is large-scale and there are a large number of active multicast sessions. An alternative is to use a *shared* multicast tree for all the sources in the group.

As illustrated in Fig. 7.6, a shared tree consists of a special router called the *core* (or the *Rendezvous Point (RP)*) of the tree, and other routers (called *on-tree* routers), which form the shortest path route from a member host's router to the core. To build a shared tree for a multicast session, a core is first chosen. Then the on-tree routers send *Join* requests to the core, and set up the routing tables accordingly. When the shared tree is set up, multicast datagrams from all the sources in the group are forwarded in this tree.

Unlike DVMRP, CBT does not broadcast the first datagram. Thus the traffic load is greatly reduced, making it suitable for multicasting in large-scale and dense networks. Moreover, the sizes of the multicast routing tables in the routers are greatly reduced, since a router only needs to store information for each multicast group, i.e., the number of CBT router entries is the same as the number of active groups. Recall that in DVMRP or MOSPF, an intermediate router needs to store information for each source in every multicast group, resulting in the DVMRP router entries of $\sum_{i \in (\text{active groups})}$ (No. of sources in group i). However, CBT has the *traffic concentration* problem, where all the source traffic may concentrate on a single link, resulting in congestion and a larger delay than multiple-tree schemes.

Protocol Independent Multicast (PIM)

Multicast routing protocols can be roughly classified into two types: source-tree based and shared-tree based. Clearly, each type has its strengths and limitations. For example, using a separate tree for each source facilitates a more even distribution of the multicast traffic in the network. Moreover, multicast datagrams from a source are distributed in the shortest path tree, resulting in a better delay performance. However, each multicast router has to maintain states for all sources in all multicast groups. This may be too costly when there are a large number of multicast sessions. Shared-tree-based protocols solve this problem by using a shared tree for all the sources in a group, resulting in a greatly reduced number of states in the routers. However, this is at the cost of the traffic concentration problem. Moreover, the shared tree may not be optimal for all the sources, resulting in larger delay and jitter. Also, the performance of the shared tree largely depends on how the Rendezvous Point is chosen.

Since a multicast protocol may be used in various situations, where the number of participants and their locations, the number of sources, and the traffic sent by each source may be highly diverse, it is very difficult to find a single protocol which is suitable for all of the scenarios. A solution to this problem is to use a *multi-modal* protocol that can switch its operation mode for different applications. The Protocol Independent Multicast protocol (PIM) is such a protocol with two modes: the *dense mode* where source-based trees are used, and the *sparse mode* where a shared tree is used. In the dense mode, PIM works like DVMRP. In the sparse mode, PIM works like CBT. When there is a high-rate source, its local router may initiate a switch to the source-based tree mode and use a source-based shortest path tree for that source.

7.2.4 The multicast backbone: MBone

MBone stands for the multicast backbone. It was created in 1992, initially used to send live IETF meetings around the world. Over the years, MBone has evolved to become a semi-permanent IP multicast testbed, consisting of volunteer participants (e.g., network providers and institutional networks).

It has been used for testing of new protocols or tools (e.g., the *vic* teleconferencing tool in 1994), live multicasting of academic conferences (e.g., ACM SIGCOMM), the NASA space shuttle missions, and even a Rolling Stones concert.

MBone is an *overlay network* with a double-layer structure. The lower layer consists of a large number of local networks that can directly support IP multicast, called multicast *islands*. The upper layer consists of a mesh of point-to-point links, or *tunnels*, connecting the islands. The **mrouted** multicast routing daemon is running at the end points of the tunnels using the DVMRP protocol. Multicast IP datagrams are sent and forwarded within the islands. However, when a multicast IP datagram is sent through a tunnel, it is encapsulated in a unicast IP datagram. When the unicast IP datagram reaches the other end of the tunnel, the unicast IP header is stripped and the recovered multicast IP datagram is forwarded. Note that such a dual-layer structure is also suggested and used in IPv6 deployment.

7.2.5 Configuring a multicast router

Configuring IGMP

IGMP is automatically enabled when a multicast protocol is started on a router interface. The following command can be used in the *Interface Configuration* mode (see Section 3.3.2) to have a router interface join a multicast group. The **no** form of this command cancels the group membership.

ip igmp join-group group-address

no ip igmp join-group group-address

The frequency at which IGMP requests are sent can be configured using the following commands in the *Interface Configuration* mode. The **no**-version of the command restores the default value of 60 seconds.

ip igmp query-interval *new-value-in-seconds*

no ip igmp query-interval

To display IGMP related router configuration or information, use the following command in the *Privileged EXEC* mode.

- **show ip igmp groups**: Displays the multicast groups in the attached networks.
- **show ip igmp interface**: Displays multicast related information on a router interface.
- **debug ip igmp**: Displays IGMP packets received and transmitted.

Configuring multicast routing

It takes two steps to set up multicast routing in a Cisco router. First, enable multicast routing using the following command in the *Global Configuration* mode. The **no**-version of the command disables multicast routing.

ip multicast-routing

no ip multicast-routing

Next, configure each router interface in the *Interface Configuration* mode, e.g., specifying which multicast routing protocol to use. The following command enables PIM on an interface and sets the mode in which PIM works:

ip pim [dense-mode | sparse-mode | dense-sparse-mode].

When **dense-sparse-mode** is specified in the above command, PIM operates in a mode determined by the group. The following commands can be used to display multicast related information in a Cisco router in the *Global Configuration mode*.

- **show ip mroute**: Displays the multicast routing table.
- **show ip mroute summary**: Displays a one-line summary for each entry in the multicast routing table.
- **show ip mroute count**: Displays multicast statistics.
- **show ip dvmrp route**: Displays the DVMRP routing table.
- **show ip pim neighbor**: Lists PIM neighbors discovered by the router.
- **show ip pim interface**: Displays router interface configurations.

Cisco IOS multicast diagnostic tools

Cisco IOS provides several multicast diagnostic tools as listed in the following. These tools are executable in the *Privileged EXEC* mode.

- **mtrace** *source [destination] [group-IP]*: Traces the path from a source to a destination in a multicast tree.
- **mrinfo** *[hostname | address] [source-address | interface]*: Verifies multicast neighbors and shows multicast neighbor router information.
- **mstat** *source [destination] [group-IP]*: Shows IP multicast paths in an ASCII graphic format, as well as statistics such as packet drops, duplicates, TTLs, and delays.
- **ping** *group-IP*: This command can be executed in both a router and a host. When a multicast group IP address is pinged, all the interfaces in the group will respond.

7.3 Realtime multimedia streaming

7.3.1 Realtime streaming

Realtime multimedia applications are increasingly popular in today's Internet. Examples of such applications are video teleconferencing, Internet telephony or Voice over IP (VoIP), Internet radio, and video streaming. These new applications raise new issues in network and protocol design.

VoIP enables telephony service, traditionally provided over circuit switched networks, e.g., the Public Switched Telephone Network (PSTN), in the packet-switched, best effort Internet. With this service, the voice signal is digitized at the source with an *analog to digital converter*, segmented into IP packets and transmitted through an IP network, and finally reassembled and reconverted to analog voice at the destination. Some of the underlying protocols used in VoIP service will be covered in this section.

Another example of realtime service is video streaming, as illustrated in Fig. 7.7. Frames are generated at the source (e.g., from the output of a video camera) continuously, and then encoded, packetized and transmitted. At the receiver, the frames are reassembled from the packet payloads and decoded. The decoded frames are then continuously displayed at the receiver. The network should guarantee delivery of the video packets at a speed matching the display rate, otherwise the display will stall.

However, the underlying IP network only provides connectionless, *best-effort* service. Video packets may be lost and delayed, or arrive at the receiver out of order. This is further illustrated in Fig. 7.8. Although the

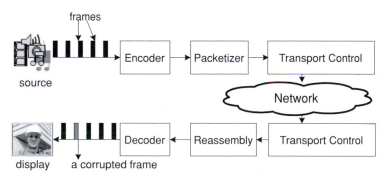

Figure 7.7. The architecture of Internet video streaming.

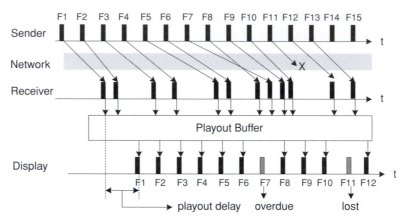

Figure 7.8. A video streaming example: the playout buffer is used to absorb jitter.

video frames are sent periodically at the source, the received video frame pattern is distorted. Usually the receiver uses a playout buffer to absorb the variation in the packet interarrival times (called *jitter*). Each frame is delayed in the playout buffer for a certain amount of time and is extracted from the buffer at the same rate at which they are transmitted at the source. An overdue frame, which arrives later than its scheduled time for extraction from the buffer (or the time it is supposed to be displayed), is useless and discarded. The difference between the arrival time of the first frame and the time it is displayed is called *playout delay*. With a larger playout delay, a frame is due at a later time, and thus a larger jitter is tolerable and fewer frames will be overdue. But this improvement in loss rate is at the cost of a larger delay experienced by the viewer.

In addition to the jitter control discussed above, there are many other requirements for effective realtime multimedia streaming. These requirements can be roughly categorized into two types: *end-to-end transport control* and *network support*. End-to-end transport control is implemented at the source and receiver, assuming a stateless core network, while network support is implemented inside the network. Several important end-to-end controls for realtime streaming are listed here.

- **Sequence numbering**. As shown in the above example, there needs to be a means for the receiver to detect if the arriving packets are out of order. One way to do this is to assign a unique identifier, called the *sequence number*, to each packet. The sequence number is increased by one for each packet transmitted. By examining the sequence numbers of the arriving packets, the receiver can tell if a packet is out-of-order or if a packet is lost.

- **Timestamping**. The source should send the *sampling instance* of each frame to the receiver, so that the receiver can replay the frames at the right pace. Timestamps can also be used by a receiver to compute jitter and round trip time.
- **Payload type identification**. As there are many different multimedia data types, coding schemes, and formats, the sender should inform the receiver about the payload type, so that the receiver can interpret the received data.
- **Error control**. Since the underlying IP network is unreliable, error control is needed to protect video packets. Traditional error control techniques include Forward Error Correction (FEC) and Automatic Repeat reQuest (ARQ).
- **Error concealment**. When packets are lost, the receiver may perform error concealment to reduce the impact of the lost packets. For example, when a frame is lost, the player may repeat the previous frame, or interpolate the lost frame using adjacent frames.
- **Quality of Service (QoS) feedback**. The receiver may collect statistics, such as loss rate, jitter, received frame quality, and send them back to the sender. With such information, the sender may adjust its parameters or operation modes to adapt to congestion or packet losses in the network.
- **Rate control**. A multimedia session may have a high data rate (e.g., high quality video streaming). Usually UDP is used for multimedia data transfer. The high-rate UDP data flows may cause congestion in the network, making other adaptive TCP flows suffer from low throughput (see Exercise 9 in Chapter 6). The sender needs to be adaptive to network congestion. When there is congestion, the sender may reduce its sending rate, e.g., by reducing the frame rate or changing the encoding parameters.

In addition to the end-to-end transport controls, realtime multimedia streaming also requires support from the packet-switched IP network. Examples of such supports are: (1) reservation of bandwidth along the network path for a multimedia session; (2) scheduling packets at the core routers to guarantee their QoS requirements; (3) sophisticated routing algorithms to find a route that satisfies the QoS requirements of a multimedia session (e.g., enough bandwidth or a low loss probability); and (4) shaping and policing the multimedia data flow to make it conform to an agreed-upon traffic specification.

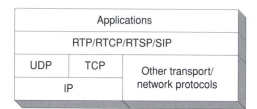

Figure 7.9. The protocol stack supporting multimedia services.

7.3.2 Protocols supporting realtime streaming services

Figure 7.9 shows the protocol stack supporting multimedia services. As shown in the figure, there are several such protocols at the application layer, e.g., the Realtime Transport Protocol (RTP), the Realtime Transport Control Protocol (RTCP), the Real Time Streaming Protocol (RTSP), and the Session Initiation Protocol (SIP). UDP is usually used at the transport layer, providing multiplexing and header error detection (checksum) services. There are a number of reasons why TCP is not used for multimedia transport. For example, the delay and jitter caused by TCP retransmission may be intolerable, TCP does not support multicast, and TCP slow-start may not be suitable for realtime transport.

RTP is an application layer transport protocol providing essential support for multimedia streaming and distributed computing. RTP encapsulates realtime data, while its companion protocol RTCP provides QoS monitoring and session control.

RTP/RTCP are application layer protocols. Usually they are integrated into applications, rather than a separate standard protocol module in the system kernel. This makes it flexible, allowing it to support various multimedia applications with different coding formats and transport requirements. RTP is deliberately not complete. A complete specification of RTP requires a set of profiles defining payload type codes, their mapping into the payload formats, and payload specifications. RTP/RTCP is independent of the underlying transport and network layer protocols. RTP/RTCP does not by itself provide timely delivery or other QoS guarantees. Rather, RTP/RTCP relies on the lower-layer protocols for reliable service. Figure 7.10 shows the RTP header format. The fields are listed here.

- Version (V): 2 bits. This field shows the RTP version, which is currently 2.
- Padding (P): 1 bit. If this bit is set to 1, the RTP payload is padded to align to the 32-bit word boundary. The last byte of the payload is the number of padding bytes.

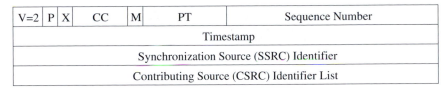

Figure 7.10. The RTP header format.

- `Extension (X)`: 1 bit. If set, there is a variable size extension header following the RTP header.
- `CSRC Count (CC)`: 4 bits. This field indicates the number of contributing source (CSRC) identifiers that follow the common header. A CSRC is a source that contributes data to the carried payload.
- `Marker (M)`: 1 bit. The interpretation of this bit is defined by a profile. This bit can be used to mark a significant event, e.g., the boundary of a video frame, in the payload.
- `Payload Type (PT)`: 7 bits. This field identifies the format of the RTP payload and determines its interpretation by the application. For example, the payload type for JPEG is 26, and the payload type for H.261[2] is 31.
- `Sequence Number`: 16 bits. This field is the sequence number of the RTP packet. The initial value of the field is randomly generated. The value is increased by 1 for each RTP packet sent. This field can be used for loss detection and resequencing.
- `Timestamp`: 32 bits. This field identifies the sampling instant of the first octet of the RTP payload, used for synchronization and jitter calculation.
- `Synchronization Source (SSRC) Identifier`: 32 bits. This field identifies the synchronization source, which is the source of a RTP packet stream.
- `Contributing Source (CSRC) Identifier List`: 0 to 15 items, each with 32 bits. The list of identifiers of the sources whose data is carried (multiplexed) in the payload.

RTCP uses several types of packets, e.g., Sender Report (SR) and Receiver Report (RR) for QoS reports, Source Description (SDES) to describe a source, goodbye (BYE) packet for leaving the group, and other application-specific packets (APP). A RTCP packet may be the concatenation of several such packets. The format of a RTCP SR packet is shown in Fig. 7.11. A RTCP RR packet has the same format as a RTCP SR, but with the PT field set to 201 and without the Sender Info block. The following list gives the definitions of the header fields.

---

[2] A video coding standard published by the International Telecommunications Union (ITU).

V=2	P	RC	PT=SR=200	Length	Header

SSRC of Sender	

NTP Timestamp, most significant word	Sender
NTP Timestamp, least significant word	Info
RTP Timestamp	
Sender's Packet Count	
Sender's Octet Count	

SSRC_1 (SSRC of First Source)		Report
Fraction Lost	Cumulative Number of Packets Lost	Block 1
Extended Highest Sequence Number Received		
Interarrival Jitter		
Last SR (LSR)		
Delay Since Last SR (DLSR)		

SSRC_2 (SSRC of second source)	Report
...	Block 2

Profile–specific Extensions

Figure 7.11. The format of a RTCP sender report.

- NTP Timestamp: 64 bits. This field carries the wallclock time (absolute time) when the report is sent. It is used in the round trip time calculation.
- Sender's Packet Count: 32 bits. The total number of RTP packets sent by this sender.
- Sender's Octet Count: 32 bits. The total number of RTP bytes sent by this sender.
- Fraction Lost: 8 bits. This field is the fraction of RTP data packet lost from the source during the last reporting period.
- Cumulative Number of Packets Lost: 24 bits. This field is the total number of RTP data packet lost from the source since the beginning of the session.
- Extended Highest Sequence Number Received: 32 bits. The lower 16 bits of this field contain the highest sequence number received in a RTP packet from the source. The higher 16 bits contain an extension of the sequence number with the corresponding count of sequence number cycles.
- Interarrival Jitter: 32 bits. This is an estimate of the statistical variance of the RTP data packet interarrival time, measured in timestamp units (e.g., sampling periods) and expressed as an unsigned integer.

- Last SR (LSR): 32 bits. This field is the middle 32 bits out of 64 in the NTP timestamp received as part of the most recent RTCP SR packet from the source.
- Delay since Last SR (DLSR): 32 bits. This field is the delay, expressed in unit of 1/65536 second, between receiving the last SR packet from the source and sending this reception report block.

During a RTP session, participants send QoS reports (SR or RR) periodically. However, the number of participants in a RTP multicast session could range from 1 to a very large number. If every participant sends QoS reports at a fixed rate, when the group size is large, there will be a large amount of QoS report traffic, eventually causing congestion and leaving no bandwidth for other data. Rather, based on the numbers of senders and receivers in the last report period, RTP calculates the report transmission interval dynamically, keeping the bandwidth used on the reports a relatively constant portion of the total bandwidth used in the RTP session. RTP also enforces a 5-second minimum interval between consecutive reports.

RTSP is an application layer control protocol for initiating and directing realtime streaming. It provides the "Internet VCR remote control" functions, e.g., *pause*, *fast-forward*, and *rewind*. RTSP can be transported using UDP or TCP. It works with RTP/RTCP for controlled streaming, as illustrated in Fig. 7.12.

SIP is another application layer control protocol providing signaling for realtime transport in the Internet. There are two basic components in SIP, i.e., the *SIP user agent* which is the end system of a call, and the *SIP network server* which handles the signaling procedures. SIP is widely used in IP telephony. Its International Telecommunication Union (ITU) counterpart is the ITU-T H.323 standard.

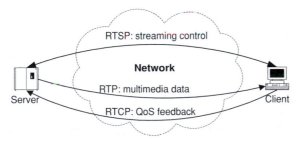

Figure 7.12. A multimedia streaming example.

7.3.3 Java Media Framework and JMStudio

RTP/RTCP was first proposed in early 1996 as RFC 1889. Since then, there have been a number of independent RTP/RTCP implementations, e.g., the *vic* video conferencing tool and the Java^{TM} Media Framework (JMF).

JMF is a Java application programming interface (API) for multimedia data processing and streaming. It provides a library package for transport and playout of realtime media. Since it is written in Java, it has the advantage of *platform independence*, i.e., once written, the same program runs on various platforms. RTP/RTCP and RTSP are all implemented in JMF.

In the JMF installation package, there is a tool called *JMStudio* which can stream video using RTP/RTCP. In the following exercises, we will use JMStudio as a tool to examine video streaming.

7.4 Simple multicast exercises

For all the exercises in this section, the network topology is given in Fig. 1.3, where all the hosts are connected to a single network segment using their default IP addresses, i.e., from 128.238.66.100 to 128.238.66.107.

Exercise 1 Execute **netstat -rn** to display the routing table of your host. If there is no entry for the 224.0.0.0 subnet, you need to provide a default route for multicast traffic, by:

route add -net 224.0.0.0 netmask 240.0.0.0 dev eth0[3]

Save the new routing table.

In Solaris, usually this entry is already in the routing table.

LAB REPORT Submit the routing table you saved.

Exercise 2 Execute **netstat -g** to show the multicast group memberships for all the interfaces in your host.

LAB REPORT How many multicast groups did the interface belong to? What were the groups? Explain the meaning of the group IDs.

---

[3] This command can be appended to the /etc/rc.local file, so that it will be executed automatically when the system bootstraps. Each time when the network interface is brought down and up again by the **ifconfig** command, you may need to run the **route** command to re-insert the multicast routing entry.

Exercise 3 | Execute **ping 224.0.0.1**. Examine the **ping** output to see which hosts reply.

Ping a broadcast address using **ping -b 128.238.66.255**. Examine the **ping** output to see which hosts reply.

In Solaris, use **ping -sv** *target_IP* in both cases.

LAB REPORT   Which hosts replied when the multicast address was pinged? Which hosts replied when the broadcast address was pinged?

In each case, was there a reply from your host?

Exercise 4 | Execute **tcpdump -n -nn -e** and **tcpdump ether multicast -n -nn -e** to capture an Ethernet unicast frame, an Ethernet multicast frame, and an Ethernet broadcast frame.

To generate an Ethernet unicast frame, run **sock -i -u -n1** *remote_host* **echo**.

Execute **sock -i -u -n1 230.11.111.10 2000** to generate an Ethernet multicast frame.

Generate another Ethernet multicast frame, but with a different group address of 232.139.111.10.

To generate an Ethernet broadcast frame, you may **ping** a remote host that has no entry in the ARP table of you host. Recall that the ARP request is broadcast.

Save the frames captured for the lab report.

LAB REPORT   Compare the source and destination MAC addresses of the frames you captured.

Use one of the multicast frames captured to explain how a multicast group address is mapped to a multicast MAC address. For the two multicast frames captured, do they have the same destination MAC address? Why?

Exercise 5 | Start the multicast client **netspy** on all the hosts, by executing

netspy 224.111.111.111 1500.

Then, start the multicast sender **netspyd** on shakti, by executing

netspyd 224.111.111.111 1500 1.

Execute **tcpdump ip multicast** on every host to capture multicast IP datagrams.

Login to shakti from a remote machine, e.g., kenchi, using **telnet** or **ssh**.

Save the captured multicast datagram sent by **netspyd** and exit the **telnet** (or **ssh**) session.

LAB REPORT   From the **tcpdump** output, how many messages are sent by **netspyd** when a new user logged in to shakti? From the **netspy** outputs on all the hosts, how many copies of the message are received in total?

Did shakti, where the multicast sender, **netspyd**, was running, receive the multicast datagram? Why? If yes, through which interface did shakti receive this datagram?

Exercise 6 | Keep the **netspy** and the **tcpdump** programs running. Execute **ping 224.111.111.111** from kenchi. Examine the **tcpdump** and **ping** outputs to see which hosts replied.

To avoid confusion, students should do this exercise by turns.

Terminate the **netspy** programs on several hosts, e.g., shakti, vayu, and fenchi. Execute the **ping** command again. Also, examine the **tcpdump** and the **ping** outputs to see which hosts replied.

7.5 IGMP exercises

In the following exercises, students are divided into two groups, Group A and Group B, each with four hosts and one router. The network topology of each group is given in Fig. 7.13, and the corresponding host IP addresses and router IP addresses are given in Table 7.2 and Table 7.3, respectively.

Table 7.2. *Hosts IP addresses for Fig. 7.13*

		GROUP_A		GROUP_B
Host	Name	IP address	Name	IP address
host1	shakti	128.238.63.100/24	yachi	128.238.64.100/24
host2	vayu	128.238.63.101/24	fenchi	128.238.64.101/24
host3	agni	128.238.64.103/24	kenchi	128.238.65.100/24
host4	apah	128.238.64.104/24	guchi	128.238.65.101/24

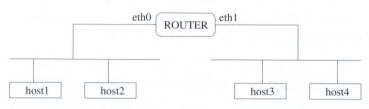

Figure 7.13. The network topology for the exercises in Section 7.5.

Table 7.3. *Router IP addresses for Fig. 7.13*

Group	Name	eth0	eth1
Group A	router3	128.238.63.3/24	128.238.64.3/24
Group B	router4	128.238.64.4/24	128.238.65.4/24

Exercise 7 Connect the hosts and the route in your group as shown in Fig. 7.13. Set the IP address of your host as given in Table 7.2. Note that the IP addresses of the router interfaces are the same as their default IP addresses.

Login to the router and run **ip multicast-routing** to enable multicast routing in the *Global Configuration* mode. Then, enable the PIM protocol on each interface, by running **ip pim dense-mode** in the *Interface Configuration* mode.[4] Now the router is enabled to do multicast routing using PIM.

Login to the router, execute **show ip igmp interface** and **show ip igmp group** in the *Privileged EXEC* mode. Examine the multicast group memberships currently recorded in the router and the configurations of the router interfaces.

Exercise 8 Start **netspy** on all the hosts, by using:

> netspy 224.111.111.111 1500.

Start **netspyd** on host1 (shakti in Group A and yachi in Group B), by using:

> netspyd 224.111.111.111 1500 16.

Login to the router. Run **show ip igmp interface** and **show ip igmp group** in the *Privileged EXEC* mode again to examine the current membership records.

Try if you can **ping** a host on the other side of the router. Login to host1 from host2 in your group, then logout. See if the multicast messages sent by **netspyd** reach the other side of the router.

LAB REPORT Can you ping a host on the other side of the router? Will the router forward a multicast IP datagram to the other side? Justify your answers.

Exercise 9 Execute **tcpdump ip multicast -w ex9a.out** in one console to capture IGMP messages. At the same time, execute **tcpdump ip multicast** in another console to monitor the capture process. When you see three or more IGMP queries in the second **tcpdump** output, terminate both **tcpdump** programs.

Start **ethereal** by using **ethereal -r ex9a.out** to analyze the IGMP messages you captured. Print and save two different IGMP messages.

---

[4] As usual, each router should be configured by one person to avoid confusion.

Repeat the above experiment, but change the output file to `ex9b.out`. Terminate **netspy** on host2 and host4. Terminate the **tcpdump** programs and analyze the IGMP *leave* message you captured.

LAB REPORT   What is the value of the Time-to-Live (TTL) field for the IGMP messages? Why do we not set the TTL to a larger number?

What is the default frequency at which the router sends IGMP queries?

Exercise 10   Login to the router. See if you can make a router interface (e.g., `Ethernet0`) join a multicast group of 224.0.0.2, using
ip igmp join-group 224.0.0.2.

LAB REPORT   Explain why the above command fails.

7.6 Multicast routing exercises

For the rest of the exercises in this chapter, the network topology is given in Fig. 7.14. The exercises will be jointly performed by all the students. The IP addresses of the hosts and router interfaces are given in Fig. 7.14.

Exercise 11   Connect the hosts and routers as illustrated in Fig. 7.14. Configure the IP addresses of the hosts and router interfaces as given in the figure. Note that most of the router interfaces use their default IP addresses, only the `Ethernet0` interface of `Router4` needs to be changed to 128.238.63.4.

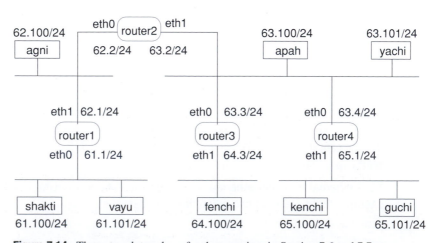

Figure 7.14. The network topology for the exercises in Section 7.6 and 7.7.

Enable PIM multicast routing in all the routers (see Exercise 7).

Run **tcpdump ip multicast** on all the hosts.

Execute **netspy 224.111.111.111 1500** on `shakti`, `agni`, `apah`, `fenchi`, and `kenchi`. Execute **netspyd 224.111.111.111 1500 16** on `yachi`. To generate multicast traffic, you can login (by **telnet** or **ssh**) to or logout of `yachi`. Each time when the login user set of `yachi` changes, **netspyd** on `yachi` will send a multicast datagram to group 224.111.111.111, to report the change in its login users.

Can you see the **netspy** messages on the 128.238.65.0 (or the 128.238.61.0) subnet in the **tcpdump** output?

Terminate the **netspy** program on `kenchi` (or `shakti`). Can you see the **netspy** messages on the 128.238.65.0 (or the 128.238.61.0) subnet?[5]

Save one of the PIM routing packets. You may use **ethereal** to analyze it.[6] What is the destination IP address used in this PIM routing packet?

LAB REPORT   Answer the above questions.

Exercise 12 | In this exercise, try the **mstat** Cisco IOS command to find the multicast tree from a source. The **mstat** command is executable in the *Privileged EXEC* mode. You can always type "?" to get help on the syntax of the command.

Exercise 13 | Keep **netspy** running on all the hosts. Ping the multicast group address from `yachi`, using

ping 224.111.111.111 -t *n*.

The parameter *n* is the TTL to be set to the multicast datagrams sent by **ping**. Try different values of *n*, e.g., 1, 2, 3, and 16. See how far a multicast datagram can travel with different TTL values.

Now, login to `Router2`, in the *Interface Configuration* mode, set the TTL threshold of the `Ethernet0` interface to 32, using:

ip multicast ttl-threshold 32.[7]

Run the **ping** command with $n = 16$ again. Can you see the multicast datagrams in the 128.238.61.0 and 128.238.62.0 subnet? Try $n = 33$. Answer the same question.

LAB REPORT   Answer the above questions.

What is the use of the TTL threshold in the router interface?

---

[5] If IGMPv1 is used, a participant does not send a *leave* message when it leaves the group. In this case, the membership record in the router expires in 120 seconds. During this interval, the router still forwards multicast datagram through the port.

[6] As usual, first capture the packet using **tcpdump -w** *output_file*, then open the output file using **ethereal**.

[7] The syntax of this command may be different for different versions of Cisco IOS. You may use "?" to get help.

7.7 Multicast video streaming exercise

In the following exercise, we use **jmstudio** for video streaming. The routers and hosts have the same configurations as in Fig. 7.14.

Exercise 14 Start **jmstudio** on all the hosts, by using **jmstudio &**.

On shakti, go to the **jmstudio** menu: File/Transmit .... In the "RTP Transmit" dialog, chose file /home/guest/video/Hurr-Lili-Trailer.mpeg. Then click the "next" button. In the next window, click the "next" button again. In the following window, specify the multicast group address to be 224.123.111.101, with port number 22224 and TTL 33. Then click the "Finish" button. Now the **jmstudio** on shakti is transmitting the video clip using RTP/RTSP/UDP/IP to the multicast group 224.123.111.101 on port 22224.

On all other hosts, go to the **jmstudio** menu: File/Open RTP Session .... In the following "Open RTP Session" dialog, specify the same group address, port number and TTL as that used in shakti. Now you should see the received video is displayed on the screen.

Execute **tcpdump ip multicast -w ex14.out** in one console to capture the multicast datagrams. In another console, execute **tcpdump ip multicast** to monitor the capture process. When you see some RTCP packets in the second **tcpdump** output, terminate both **tcpdump** programs.

Use **ethereal** to load the ex14.out file. Analyze the header format of a RTP data packet and a RTCP Sender (or Receiver) Report packet.

8 The Web, DHCP, NTP and NAT

The dream behind the Web is of a common information space in which we communicate by sharing information.   Tim Berners-Lee

8.1 Objectives

- The HyperText Transfer Protocol and the Apache web server.
- The Common Gateway Interface.
- The Dynamic Host Configuration Protocol.
- The Network Time Protocol.
- The Network Address Translator and the Port Address Translator.
- An introduction to socket programming.

8.2 The HyperText Transfer Protocol

8.2.1 The HyperText Transfer Protocol and the Web

In the early days of the Internet, email, FTP, and remote login were the most popular applications. The first *World Wide Web* (WWW) browser was written by Tim Berners-Lee in 1990. Since then, WWW has become the second "Killer App" after email. Its popularity resulted in the exponential growth of the Internet.

In WWW, information is typically provided as HyperText Markup Language (HTML) files (called *web pages*). WWW resources are specified by Uniform Resource Locators (URL), each consisting of a protocol name (e.g., http, rtp, rtsp), a ":://", a server domain name or server IP address, and a path to a resource (an HTML file or a *CGI script* (see Section 8.2.2)). The HyperText Transfer Protocol (HTTP) is an application layer protocol

Table 8.1. *A HTTP client request and an HTTP server response*

A HTTP client request	GET /usage/try1.htm HTTP/1.1\r\n Accept: image/gif, image/jpeg, */*\r\n Accept-Language: en-us\r\n Accept-Encoding: gzip, deflate\r\n User-Agent: Mozilla/4.0 (Red Hat Linux)\r\n Host: 128.238.42.129\r\n Connection: Keep-Alive\r\n \r\n
A HTTP server response	HTTP/1.1 200 OK\r\n Date: Sat, 18 Oct 2003 19:28:32 GMT\r\n Server: Apache/2.0.40 (Red Hat Linux)\r\n Last-Modified: Sat, 18 Oct 2003 04:11:58 GMT\r\n Accept-Ranges: bytes\r\n Content-Length: 529\r\n Connection: close\r\n Content-Type: text/html; charset=ISO-8859-1\r\n \r\n Data (529 bytes)

for distributing information in the WWW. In common with many other Internet applications, HTTP is based on the client–server architecture. An HTTP server, or a *web server*, uses the well-known port number 80, while an HTTP client is also called a *web browser*. With HTTP, a browser sends a request to a server for a file, and the server responds with the requested file if it is available.

HTTP messages are *English-based* and flexible, which are different from those lower layer protocols where the messages are *code-based*. As the examples in Table 8.1 show, an HTTP message consists of the following components.

- A start-line. In an HTTP request, usually the start-line is a request for a file. In the example, "GET" is an HTTP *method* which instructs the server to return a file. The next parameter is the directory and name of the target file. The last parameter is the version of HTTP. There are two versions, HTTP/1.0 and HTTP/1.1, that are widely used in the WWW. In an HTTP response, the start-line contains a code indicating the status of the corresponding request. The definitions of the codes are given in Table 8.2.

Table 8.2. *Definitions of the HTTP response status codes*

Codes	Meaning	Example
1XX	Informational	100: The server has received the first part of the request.
2XX	Success	200: The request is successful and the response is returned in the following message body.
3XX	Redirection	300: Multiple choices.
4XX	Client error	404: The requested file is not found.
5XX	Server error	500: Internal server error.

- Optional headers. Following the start-line is a list of HTTP headers. Each line consists of a header name and a value. These headers are English-based with intuitive meanings. For example, the header "Host: 128.238.42.129\r\n" means that this request is from host 128.238.42.129. The header "User-Agent: Mozilla/4.0 (Red Hat Linux)\r\n" means the client is using the Red Hat Linux operating system and the web browser is Mozilla 4.0. Headers are optional. Each header line ends with a "\r\n".
- A blank line, i.e., a "\r\n" only.
- For an HTTP response, the requested file or other data follows the blank line.

HTTP uses TCP for file transfers. Before an HTTP request, the client first establishes a TCP connection to the server. After the HTTP response is sent, the server may terminate the TCP connection. Many HTML files have embedded objects, e.g., pictures, audio/video, or Java *applets*. When an HTML file is received, the web browser parses the file to identify the embedded objects, and then sends an HTTP request to the server for each embedded object. For example, for an HTML file with two pictures, the client sends three HTTP requests, with the first one for the original file and the other two for the pictures. With HTTP/1.0, the client establishes a TCP connection for each request. The TCP connection is terminated by the server when transmission of the requested file (an HTML file or an embedded object) is over. This may be inefficient when the HTML file has many embedded objects, since TCP connection establishments and terminations waste both network and server resources and introduce additional delays. In HTTP/1.1, *persistent connection*s are supported, where all the embedded objects are sent through the TCP connection established for the first request. In the example in Table 8.1, the persistent connection feature is enabled at the client (see header "Connection: Keep-Alive"), and disabled at the server (see header "Connection: close"). Note that a TCP connection is persistent if and only if both the client and server enable this feature.

HTTP allows the use of *proxies* for web access control and better performance. A proxy is an intermediary between an HTTP client and an HTTP server. To the client, the proxy acts as if it were the target server, while to the server, the proxy acts as a normal client. When a proxy is used for a local network, local clients direct all their HTTP requests (to various servers) to the proxy. A request to the proxy contains the full URL in the start-line, rather than just the directory and the target file name in a normal HTTP request (Table 8.1), in order to inform the proxy server which server to forward the request to. Proxies can be used with *firewalls* to block undesired traffic. Furthermore, a proxy usually maintains a web cache for recently downloaded files. Web caching exploits the property of *temporal locality* in HTTP requests.[1] When a request arrives, the proxy first searches the cache. If there is a hit, the cached file is returned to the client and there is no need to request the file from the remote server. With web caching, both the response time experienced by the client and the work load at the web server are reduced.

8.2.2 The Common Gateway Interface procotol

Web pages used to be static, consisting of text and pictures. With Java technology and the Common Gateway Interface (CGI) protocol, web pages can now be dynamic. A user can send data to the server and web pages can be generated on-demand. Dynamic web pages provide a two-way communication between web clients and web servers, making web applications such as an on-line opinion poll or e-commerce possible.

CGI uses two files, an HTML form where a user can input data, and a CGI script that processes user input data and generates a response dynamically. Any program, written in any language, that can read input from the *standard input* (STDIN) and write output to the *standard output* (STDOUT) can be used as CGI script. With CGI, a user first downloads the HTML form using a web browser. The form consists of text inputs, checklists, and buttons. A user can make choices, input text, and click the button to submit the data. The web server then invokes the CGI script that uses the user data

---

[1] This means when a file is downloaded, it is likely to be downloaded again in the near future, by the same or a different user.

as input to generate a response (usually an HTML file) dynamically. The web server then returns the CGI response to the client.

8.2.3 The Apache web server

WWW service is provided by web servers. According to the web server survey from Netcraft, the Apache server is the most popular web server in the Internet.[2] The Apache server is an open source software, included in both the Red Hat Linux 9 and Solaris 8 installation CDs.

The Apache server daemon is **httpd**. The following command starts, stops, or restarts the Apache server:

/etc/rc.d/init.d/httpd start|stop|restart.

To check the status of the Apache server, execute **pgrep httpd**. Then the process ids of the active **httpd** processes will be displayed. To check if the web server is correctly installed and started, you can start the Mozilla browser on the server host and enter the http://localhost URL. If the test page appears in the browser window, the server is up and working properly.

The Apache server uses the /etc/httpd/conf/httpd.conf configuration file. There are more than two hundred configuration *directives* in the Apache configuration language. In most of the cases, the default values should be sufficient. The ServerName *server_name*|*server_IP* directive specifies the web server, while the Port *n* directive specifies the server port number (80 by default). The KeepAlive *on*|*off* directive turns on (off) the persistent connection feature. The root directory of the HTML files can be specified by the Document Root directory directive. By default, all HTML files are stored in the /var/www/html/ directory. The URL for the file /var/www/html/abc.html is http://*server*/abc.html. If the UserDir public_html directive is used, a user, e.g., guest, can put his/her HTML files in the /home/guest/public_html directory, which can be accessed with the http://*server*/~*guest*/*file_name* URL.

Stability, scalability, and response time are three important performance metrics in web server design. Apache is a process-based web server. When started, the initial Apache process (called the *master server*) launches one or more child processes, each listening on the HTTP port (80 by default) and handling client requests. This approach has several advantages. First, the web server is more stable. If one child process crashes, other child

---

[2] According to a survey at http://news.netcraft.com/archives/web_server_survey. htm, 64.61% of the 43,700,759 websites polled use the Apache server, as of October 2003.

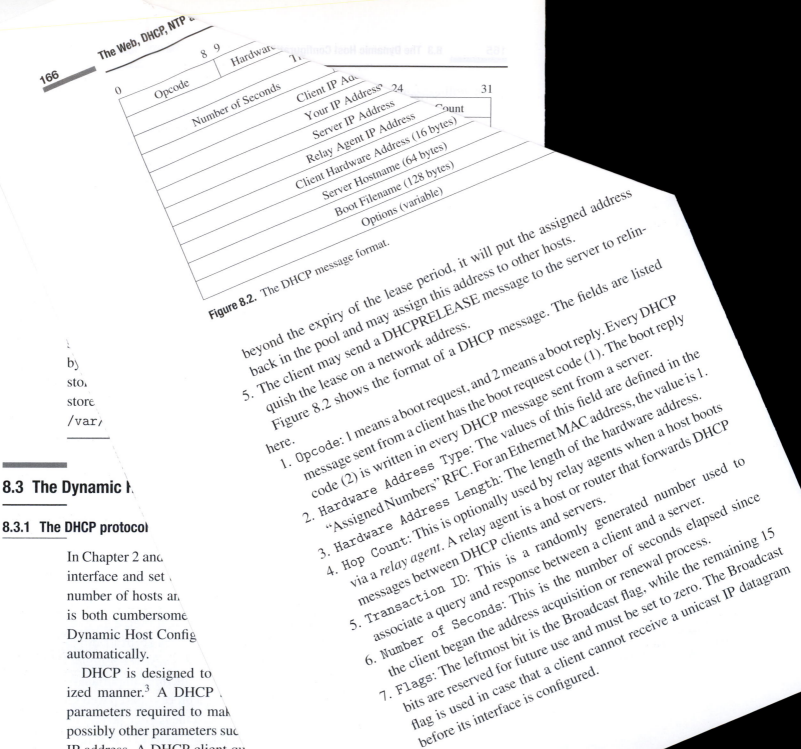

0		8	9	Hardware...				24		31
Opcode										Count

Number of Seconds

Client IP Address

Your IP Address

Server IP Address

Relay Agent IP Address

Client Hardware Address (16 bytes)

Server Hostname (64 bytes)

Boot Filename (128 bytes)

Options (variable)

Figure 8.2. The DHCP message format.

beyond the expiry of the lease period, it will put the assigned address back in the pool and may assign this address to other hosts.

5. The client may send a DHCPRELEASE message to the server to relinquish the lease on a network address.

Figure 8.2 shows the format of a DHCP message. The fields are listed here.

1. Opcode: 1 means a boot request, and 2 means a boot reply. Every DHCP message sent from a client has the boot request code (1). The boot reply code (2) is written in every DHCP message sent from a server.

2. Hardware Address Type: The values of this field are defined in the "Assigned Numbers" RFC. For an Ethernet MAC address, the value is 1.

3. Hardware Address Length: The length of the hardware address.

4. Hop Count: This is optionally used by relay agents when a host boots via a *relay agent*. A relay agent is a host or router that forwards DHCP messages between DHCP clients and servers.

5. Transaction ID: This is a randomly generated number used to associate a query and response between a client and a server.

6. Number of Seconds: This is the number of seconds elapsed since the client began the address acquisition or renewal process.

7. Flags: The leftmost bit is the Broadcast flag, while the remaining 15 bits are reserved for future use and must be set to zero. The Broadcast flag is used in case that a client cannot receive a unicast IP datagram before its interface is configured.

8.3 The Dynamic H...

8.3.1 The DHCP protocol

In Chapter 2 and...
interface and set...
number of hosts a...
is both cumbersome...
Dynamic Host Config...
automatically.

DHCP is designed to...
ized manner.[3] A DHCP...
parameters required to ma...
possibly other parameters su...
IP address. A DHCP client qu...
rameters. A DHCP server, when...
to the client. DHCP can provide...

_____

[3] An older protocol for this purpose is BOOTP, w...

8. `Client IP Address`: This field is only used when the client is in BOUND, RENEW, and REBINDING state and can respond to ARP requests.
9. `Your IP Address`: This field is the client's IP address.
10. `Server IP Address`: This is the IP address of the next server to use in the bootstrap process.
11. `Relay Agent IP Address`: This is used when booting via a relay agent.
12. `Client Hardware Address`: This is the hardware address of the client. In the case of Ethernet, the first 6 bytes are filled with the client's Ethernet address, and the remaining bytes are set to 0.
13. `Server Hostname`: This is the host name of the DHCP server.
14. `Boot Filename`: This is set in a DHCPOFFER message. The server can fill in this field with the fully qualified, null terminated path name of a file to bootstrap from.
15. `Options`: This is the optional parameter field. The `Message Type` option defines the type of the DHCP messages which is more specific than the `Opcode` field. This option has to be present in every DHCP message. Different message types are used at different stages of the client/server interaction.

8.3.2 Configuring DHCP

The DHCP server daemon is **dhcpd**, while the DHCP client daemon is **dhcpcd**. To start the DHCP server, run **/usr/sbin/dhcpd**. To start the DHCP client, run **/usr/sbin/dhcpcd**. Both commands may be appended to the `/etc/rc.d/rc.local` file to be executed automatically when the system bootstraps.

The DHCP server **dhcpd** uses a configuration file `/etc/dhcpd.conf`. Table 8.3 gives an example of the configuration file. Lines 1 and 2 define the lease time in seconds. Line 3 to line 7 define the parameters required to configure a network interface. Line 9 to line 11 define an IP address pool with two IP addresses. Multiple ranges of IP addresses can be defined here. These IP addresses can be assigned to a requesting host. Lines 13 to 16 define a static assignment, where host apah will always be assigned with a fixed IP address of 128.238.66.100. A static assignment is useful when a host is used as an Internet server, e.g., email server or web server, and a fixed IP address is required.

Table 8.3. *A DHCP server configuration file*

```
1    default-lease-time 600;
2    max-lease-time 7200;
3    option subnet-mask 255.255.255.0;
4    option broadcast-address 128.238.66.255;
5    option routers 128.238.66.1;
6    #option domain-name-servers 128.238.2.38, 128.238.3.21;
7    #option domain-name "poly.edu";
8
9    subnet 128.238.66.0 netmask 255.255.255.0 {
10      range 128.238.66.111 128.238.66.112;
11   }
12
13   host apah {
14      hardware ethernet 08:00:20:79:e9:9f;
15      fixed-address 128.238.66.110;
16   }
```

In Solaris, a DHCP server stores client configuration in two types of files: a dhcptab file that stores all the information that a client can obtain from the server, and one or more network tables mapping client identifiers to IP addresses and the configuration parameters associated with each IP address. These files can be edited by the graphical configuration tool **DHCP Manager**, or by the command line tools **dhcpconfig** (generating the table files), **dhtadm** (configuring dhcptab), and **pntadm** (managing the network tables).

The DHCP server daemon is **in.dhcpd**. The DHCP client daemon is **dhcpagent**. If the startup script finds a /etc/dhcp.*interface* file, it starts the **dhcpagent** daemon and contacts DHCP servers for configuration parameters for that interface. The Solaris DHCP client can be controlled by the **ifconfig** command: **ifconfig** *interface* **dhcp** *options*. The options are:

start: Restarts the DHCP client.

inform: Requests network information only.

extend: Requests a lease extension.

release: Releases the IP address.

drop: Drops the lease without informing the DHCP server.

status: Displays the network interface status.

8.4 The Network Time Protocol

8.4.1 The NTP protocol

With the fast growth of the Internet, accurate timing is becoming more and more important in network design, management, security, and diagnosis. For example, many network systems log network events. If the timing in a system is not accurate or the systems are not synchronized, it would be very difficult to analyze the logfiles. Some diagnostic tools, such as **tcpdump**, record packets captured along with the time when they were captured. Network measurement (e.g., DBS in Chapter 6) also requires accurate timing and synchronization.

The Network Time Protocol (NTP) is a protocol used to provide accurate timing and to synchronize computers and other network devices. NTP is an application layer protocol using UDP and TCP port 123.

Timing service is provided in the Internet in a hierarchical manner, as shown in Fig. 8.3. The NTP servers and clients are organized into 16 *strata*. An NTP primary server, or *stratum 1*, is a host synchronized with a high precision clock, e.g., an atomic clock or Global Positioning System (GPS) signals. Each server chooses one or more higher stratum servers and synchronizes with them. Choosing multiple higher stratum servers results in better reliability, since one or more of the servers in use may be down or unreachable, or their timing information may be inaccurate. The further a computer is from stratum 1, the less accurate its clock is. A list of public

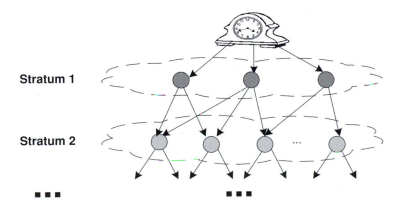

Figure 8.3. The hierarchical strata structure of NTP time service.

NTP primary time servers (stratum 1) and secondary servers (stratum 2) can be found at `http://www.ntp.org`.

In NTP, clients and servers can operate in the multicast or broadcast mode, where timing information is broadcast or multicast by the servers. On the other hand, a client can proactively poll the servers for timing information. An NTP client can synchronize with an NTP server in two ways. The client can use **rdate** or **ntpdate** to query time information from and synchronize to a remote NTP server whenever it wishes to. Furthermore, the client can start the **ntpd** daemon which synchronizes with the NTP server continuously and automatically.

8.4.2 Configuring NTP

The NTP network daemon is **ntpd**, which can be controlled by the following.

- **/etc/init.d/ntpd start|stop|restart**: Starts, stops, or restarts the NTP daemon.
- **/etc/init.d/ntpd status**: Shows the status, e.g., stopped or started, of the NTP daemon.

ntpd uses the `/etc/ntp.conf` configuration file and can be started as an NTP server or an NTP client. Some frequently used entries in the `/etc/ntp.conf` are listed here.

- **server** *server_IP|server_domain_name* [**prefer**]: Specifies an NTP server to synchronize to.
- **peer** *host_IP|host_domain_name*: Specifies an NTP peer. **ntpd** can operate in the *peer mode*, where multiple peers communicate with each other to determine which one has the most accurate clock and synchronize to it.
- **broadcastclient**: Configures a broadcast client, which receives broadcast time information to synchronize with a broadcast server.
- **multicastclient**: Configures a multicast client.
- **broadcast** *broadcast_address|multicast_address* **ttl** *ttl_value*: Configures either a broadcast or a multicast NTP server.
- **restrict** *network_address|host_IP_address* [*flags*]: Restricts NTP service to this server.

NTP also uses several other configuration files in the `/etc/ntp` directory. For example, the `/etc/ntp/ntp.drift` file contains the latest estimation of the clock frequency error, called *drift*. **ntpd** keeps on monitoring the frequency error and records the measured error in this file at hourly intervals. **ntpd** also supports authentication using the `/etc/ntp/keys` file, which stores keys and key IDs. The client and the server must have the same key and key ID in their `/etc/ntp/keys` files in order to communicate with each other. These configuration files must be specified in the `/etc/ntp.conf` file.

In Solaris, the NTP configuration files are stored in the `/etc/inet/` directory. A server configuration template `/etc/inet/ntp.server` and a client configuration template `/etc/inet/ntp.client` are provided.

8.4.3 Network timing tools

Most Linux or Unix systems provide the following NTP tools for system synchronization or other timing related tasks.

- **date**: Displays or sets the current system time.
- **rdate** [*options*] *remote_host*: Gets time from the network. The options follow.

  -P : Prints the returned time.

  -s: Sets the current system time to the returned time.

  -u: Uses UDP rather than TCP in the transport layer.

  rdate uses the RFC 868 time server,[4] with UDP and TCP port 37.
- **ntptrace** *host*: Traces time information back to the high stratum servers in the synchronization tree.
- **ntpdate** [*options*] *ntp_server*: Synchronizes the local clock with the remote server once.
- **ntpq**: Queries the state of the NTP daemon on a local or remote host.

A Cisco router can also be configured to use NTP. For example, the *Global Configuration* command **ntp server** *NTP_server_IP* specifies an NTP server with which the router will synchronize to. The *Global Configuration* command **ntp access-group** controls access to NTP service on the system.

---

[4] The *time* and *time-udp* services should be enabled in the Linux machine in order to respond to a **rdate** query, by **chkconfig time on** and **chkconfig time-udp on**, respectively.

    In Solaris, the lines corresponding to *time* and *time-udp* services in the `/etc/inetd.conf` file should be uncommented.

only the public IP addresses (which were assigned by the original ISP) need to be changed. The internal settings of the private network can remain unchanged. This is useful for large private networks where the configuration task would be time-consuming and error prone.

The disadvantage of using NAT/PAT is that the stub router may be congested and become the performance bottleneck, since in addition to IP address and port translations, the router has to recalculate the header checksums. Furthermore, ICMP error messages need to be handled carefully and the ICMP payload may also have to be translated. NAT does not directly support applications with interdependent control and data connections, e.g., H.323, RTP/RTCP, and FTP. Special *application gateways* are required to support such applications.

8.5.2 Configuring an NAT router

To configure an NAT router, do the following.

1. To specify the public IP address pool ranging from *first_IP* to *last_IP*, use the following *Global Configuration* command:

   ip nat pool *name_of_pool first_IP last_IP* **netmask** *mask*

2. To define an access list controlling which internal hosts can use the IP addresses in the pool, use the following *Global Configuration* command:

   access-list *access-list_number* **deny host** *denied_host_IP*

   access-list *access-list_number* **permit** *network_address bit_mask*

   The *access-list_number* parameter in the above commands represents an *IP standard access-list*, with valid values ranging from 0 to 99. The *bit_mask* parameter specifies which bits in the network address should be ignored. A "1" ("0") in the *bit_mask* means the corresponding network address bit should be ignored (compared).

3. Associate the access-list with the public IP address pool:

   ip nat inside source list *access-list_number* **pool** *name_of_pool*.

4. To specify a router interface which has a public IP address and connects to the Internet, use the following *Interface Configuration* commands:

   interface *name_of_interface*

   ip address *public_IP_address netmask*

   ip nat outside

5. To specify a router interface which has a private IP address and connects to the private network, use the following *Interface Configuration* commands:

interface *name_of_interface*
ip address *private_IP_address netmask*
ip nat inside

6. To define a static translation, use:

    ip nat inside source static *private_IP_address public_IP_address*
    Note that if a static translation is defined, the internal host with the *private_IP_address* should be denied from using the shared public address pool.

7. To configure PAT, use:

    ip nat inside source list *list_number* **interface** \ *router_interface*
        overload
    Then all the internal hosts use the same public IP address, i.e., the IP address of the outside router interface, using port translations.

8.6 Socket programming in a nutshell

Most of the applications discussed so far are implemented using the *socket* Application Programming Interface (API). In this section, we will give a brief overview of socket programming basics. For a more complete treatment, see Stevens [12].

As shown in Fig. 8.5, the TCP/IP protocols are implemented in the system kernel. User applications can use the TCP/IP service through the *socket* API. In such applications, each participating process should create a socket, containing the IP address of the host where the process is running

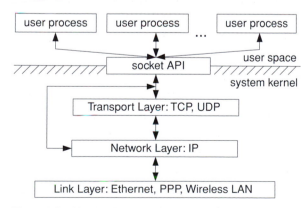

Figure 8.5. The socket API provides an interface for the user processes to access the TCP/IP services in the system kernel.

and a unique port number. Then, the application process can use the socket functions for sending or receiving data. There are three types of sockets for applications to use. If reliable transport service is required, *TCP sockets* can be used to create a TCP connection between the communicating parties. Otherwise, *UDP sockets* can be used to provide datagram service. In addition, applications can also bypass the transport layer protocols by invoking *raw sockets* to use the IP datagram service directly.

The client–server architecture is used in socket programming. A server socket listens at a port, receives client requests, and provides the requested service, while a client socket connects to the server socket to get the desired service. In the lab exercise, we will write server and client programs using TCP and UDP sockets. Some frequently used socket functions are listed below.

- **socket**(). Creates a new socket. You can specify which type of socket to create, e.g., for TCP sockets use type SOCK_STREAM, while for UDP sockets use type SOCK_DGRAM.
- **bind**(). Establishes the local association of a socket by assigning a local name to the unnamed socket.
- **setsockopt**(). Sets the current value for a socket option.
- **listen**(). Makes the TCP server socket wait for incoming requests from the TCP clients.
- **connect**(). Initiate a connection on a socket.
- **accept**(). Accepts an incoming TCP connection request.
- **send**(), **sendto**(). Transmits application data to the remote socket(s).
- **recv**(), **recvfrom**(). Receives data from the network.
- **close**(). Shuts down the socket, i.e., terminates the connection and releases the resources.

The syntax of the functions, e.g., **sendto**(), can be found in the manual pages, e.g., by typing **man sendto**, or from Stevens [12].

Figure 8.6 illustrates the typical flow of TCP and UDP socket operations. First, the user process creates a socket by calling the `socket()` function and specifying the type of socket to create, i.e., a stream socket (TCP) or a datagram socket (UDP). Then, the created socket can be assigned with a port number and IP address by calling the `bind()` function. Usually a server socket has a static IP address and a well-known port number, while `bind()` is optional for a client socket since the system will choose a random port number for the client if `bind()` is not called. If stream sockets are used, the TCP server socket then enters a listening state by calling the `listen()`

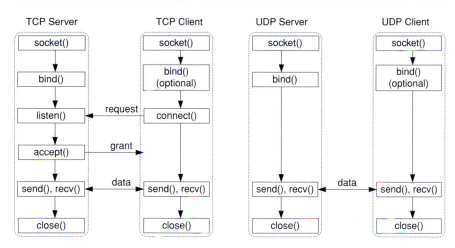

Figure 8.6. Typical flow of TCP and UDP socket operations.

function, waiting for requests from TCP clients. A TCP client, on the other hand, calls the `connect()` function to send a TCP connection request (a SYN segment) to the TCP server. When the server receives the request, it calls the `accept()` funtion to create a new socket locally, which serves as the end point of the TCP connection on the server side. In other words, the TCP server socket serves as a front-end, receiving incoming client requests and creating the corresponding local sockets. Then, data can be transmitted between the newly created socket on the server side and the TCP client socket on the client side. If datagram sockets are used, there is no need to set up or terminate the connection since UDP is connectionless. Thus, after creating the sockets and calling `bind()`, the UDP client and UDP server can directly exchange data by calling the `send()` or `recv()` functions. When data transmission is over, both the server and the client can call the `close()` function to terminate the connection (in the stream socket case) and release the resources (e.g., port numbers and memory, in both the stream and datagram socket cases).

The best way to learn programming, perhaps, is to read the source code and try it out. Four examples of socket programs, namely `UDPserver.c`, `UDPclient.c`, `TCPserver.c`, and `TCPclient.c`, are given in Appendix C.4. Moreover, the **netspy** and **netspyd** programs used in Chapter 7 are multicast socket programs and the source code is given in Appendix C.2.

To compile a socket program, e.g., TCPclient.c, use:

gcc -o TCPclient TCPclient.c -lnsl

Use the following command when compiling a socket program, e.g., TCPclient.c, in Solaris:

gcc -o TCPclient TCPclient.c -lnsl -lsocket -lresolv.

8.7 HTTP exercises

For the exercises in this section, the network topology is given in Fig. 1.3, where all the hosts are connected to a single network segment using their default IP addresses, i.e., from 128.238.66.100 to 128.238.66.107.

Exercise 1 Study the Apache server configuration file (see Section 8.2.3). Examine the various configuration directives used and the corresponding settings.

Start the Apache server on your host. In order to check if the server is working properly, you may start a `Mozilla` web browser to download the test page at `http://localhost/`.

Then, execute **pgrep httpd** to list the process IDs of the **httpd** processes started. Save the output and the configuration file for the lab report.

LAB REPORT   How many **httpd** processes were started? Which one was the master server, and which ones were the child servers? Justify your answer using the `httpd.conf` file.

What is the purpose of initiating multiple **httpd** processes?

Exercise 2 Execute **tcpdump host** *your_host* **and** *remote_host* **-w ex3.out** to capture packets between your host and a remote host.

Login to the remote host's web server: **telnet** *remote_host* **80**.

In the login console, type the following HTTP request line by line:

```
GET /usage/index.html HTTP/1.0
From: guest@your_host
User-Agent: HTTPTool/1.0
```

Note that you need to type the `Return` key to input the last line, which is blank. When the **telnet** process is terminated, save the output for your lab report.

Terminate **tcpdump**. Use **ethereal** to load the `ex3.out` file and analyze the captured HTTP packets. Print and save the HTTP request and response.

Save the HTTP response's data part into a file, named `ex3.html`. Use `Mozilla` to view the file.

LAB REPORT   Submit the HTTP request and response, including the start-lines and all the headers.

Exercise 3   By default, Apache server supports persistent connections. Before this exercise, the lab instructor should check the `KeepAlive` directive in the server configuration file to make sure it is turned on, as `KeepAlive on`.

Execute **tcpdump host** *your_host* **and** *remote_host* **-w ex4a.out** to capture packets between your host and a remote host.

Start `Mozilla` on your host. Go to menu `Edit/Preferences /Advanced/HTTP Networking`, and uncheck the `Enable Keep-Alive` checkbox to disable persistent connections.

Enter the URL `http://`*remote_host*`/try1.html`, to download the HTML file consisting a line of text, an embedded picture, and a hyperlink.

Use **ethereal** to load the `ex4a.out` file, and print the HTTP requests and responses for the lab report.

Restart the **tcpdump** program, but dump the output to a `ex4b.out` file.

Go to `Mozilla` menu `Edit/Preferences/Advanced/HTTP Networking`, and enable persistent connections by checking `Enable Keep-Alive`.

Use `Mozilla` to reload the `try1.html` file.

Use **ethereal** to load the `ex4b.out` file, and print the HTTP requests and responses for the lab report.

LAB REPORT   When you browsed the `try1.html` file for the first time, how many HTTP requests were sent? Which files were requested? How many TCP connections were used?

Answer the above questions for when you browsed the `try1.html` file for the second time.

What is the purpose of using persistent connections?

Exercise 4   Execute **tcpdump host** *your_host* **and** *remote_host* **-w ex5.out** to capture packets between your host and a remote host.

Use `Mozilla` to download the `http://`*remote_host*`/try2.htm` file, which is an HTML form, from the remote host.

Fill a text string, e.g., the name of the host being used, into the text field in the form and click the submit button in the form.

When the server response is received, terminate **tcpdump**.

Use **ethereal** to load the `ex5.out` file. Examine how CGI works, and identify the data string sent to the server. Save the HTTP request containing the data string for lab report.

LAB REPORT   Submit the data string sent to the server.

8.8 DHCP exercises

For the exercises in this section, we use the same network setting as the one used in the previous exercises.

Exercise 5   In this exercise, we use `guchi` as the DHCP server, with a configuration file shown in Table 8.3. Do the following.

1. Start the DHCP server on `guchi` in the foreground and working in the debugging mode: **/usr/sbin/dhcpd -d -f**.
2. Execute **tcpdump -exn -nn -s 100 -w exdhcp.out** to capture the DHCP messages in the network segment.
3. Then do the following to enable DHCP for the Ethernet interface on `shakti`. Go to the system menu: `System Settings/Network`. In the `Network Configuration` dialog, choose the `Device` tab, and click on the `eth0` item. Next, click the `Edit` button to bring up the `Ethernet Device` dialog. In this dialog, check `Automatically obtain IP address settings with` and select `dhcp` from the following drop list. When the configuration is done, save the new configuration and then execute **/etc/init.d/network restart** to load the new configuration.

   When `shakti` is successfully reconfigured, execute **ifconfig -a** to display its network interface configurations and execute **netstat -rn** to display its routing table. Save the outputs for the lab report.
4. Then, repeat 3 for `vayu`.
5. Repeat 3 for `agni`.
6. Repeat 3 for `apah`.

Terminate **tcpdump**. Use **ethereal** to load the `exdhcp.out` file. Print out the DHCP messages for the lab report.

Save the DHCP server output on `guchi` for the lab report.

LAB REPORT   Compare the DHCP operation captured by **tcpdump** and that shown by the DHCP server output. Explain how DHCP works.

Did `shakti` and `vayu` successfully obtain a set of new parameters? Compare the **ifconfig** and **netstat** output with the parameters carried in the corresponding DHCP messages.

Answer the above question for `agni`. Explain why `agni` failed.

Answer the above question for `apah`. Explain why `apah` succeeded.

If Solaris is used, execute **/usr/sadm/admin/bin/dhcpmgr &** to start the graphical DHCP Manager on `guchi` and configure the DHCP tables as shown in Table 8.3. Then the DHCP server can be started or stopped by **/etc/init.d/dhcp start|stop**. The DHCP queries can be generated by the **ifconfig interface dhcp start** command.

8.9 NTP exercises

Before proceeding to the next exercise, reboot the hosts to restore their original configurations.

Exercise 6 Execute **date** to display the system time of your host. Display the manual page of **date**, and study its options and usages.

Try the following **date** commands:
date −−date='2 days ago'
date −−date='3 months 2 days'
date −−set='+3 minutes'
date −r *file_name*
You can choose any file in the current directory for the *file_name* parameter. Save the outputs of the above commands.

LAB REPORT   Submit the **date** outputs you saved. Explain the use of the commands.

Exercise 7 While **tcpdump -n -nn -ex host** *your_host* **and** *remote_host* is running, execute **rdate -p** *remote_host* to display the system time of the remote machine.

Repeat the above **rdate** command, but use the **-u** option.

Save the **tcpdump** outputs for the lab report.

LAB REPORT   What port numbers were used by the remote machine? What port numbers were used by the local host?

How many bytes of data were returned by the remote time server, both in the UDP case and in the TCP case?

What TCP header options were used?

Exercise 8 In this exercise, we start the NTP server daemon on `shakti` and use NTP to synchronize all the other hosts to `shakti`.

Study the NTP configuration file `/etc/ntp.conf` in `shakti` and in your host. If you are using another machine, you can **telnet** to `shakti` and display the `/etc/ntp.conf` file in the **telnet** window.

Start the NTP server on `shakti` by: **/etc/init.d/ntpd start**. To determine the status of the NTP server, use **/etc/init.d/ntpd status**.

Use **tcpdump -ex -n -nn host** *your_host* **and shakti** to capture packets between your host and `shakti`.

Execute **ntpdate -d -v 128.238.66.100** to synchronize your host to `shakti`. Study the output of this command.

Save the **ntpdate** and the **tcpdump** outputs for the lab report.

LAB REPORT   Which port does the NTP server use? Justify your answer using the **tcpdump** output.

Exercise 9 Keep the NTP server running on `shakti`. Execute **tcpdump -exn -nn host** *your_host* **and 128.238.66.100 -w ex9.out** to capture the NTP messages between your host and `shakti`.

Start the NTP clients on your host, by **/etc/init.d/ntpd start**.

Wait for several minutes. Then terminate the **tcpdump** program. Use **ethereal** to load the `ex9.out` file. Analyze the captured NTP packets. Print one of the NTP packets for the lab report.

Execute **ntptrace** to show the client/server relation of NTP.

LAB REPORT   Submit the NTP packet captured. List the fields and their values.

What was the rate at which NTP queries were sent by the client?

Which stratum was your host in? Which stratum was the NTP server in?

8.10 NAT exercises

For the exercises in this section, we use a network setting as shown in Fig. 8.7. The lower subnet is a private network where the hosts are assigned with the Class A addresses with the 10.0.0.0/8 prefix. The upper subnet represents the Internet. The hosts, i.e., `shakti` and `vayu` are assigned with public IP addresses. `Router 1` is used as the stub router, which performs address or port translation for the private network.

Exercise 10 Connect the hosts and `Router 1` as shown in Fig. 8.7. Then set the IP address and the network mask of your host as shown in the figure. In addition, you need to

Table 8.5. *NAT Router Configuration in Fig. 8.7*

ip nat pool mypool 128.238.61.102 128.238.61.103 netmask 255.255.255.0
ip nat inside source list 8 pool mypool

ip nat inside source static 10.0.0.7 128.238.61.104

interface ethernet 0
ip address 128.238.61.1 255.255.255.0
ip nat outside

interface ethernet 1
ip address 10.0.0.1 255.0.0.0
ip nat inside

access-list 8 deny host 10.0.0.7
access-list 8 permit 10.0.0.0 0.0.0.255

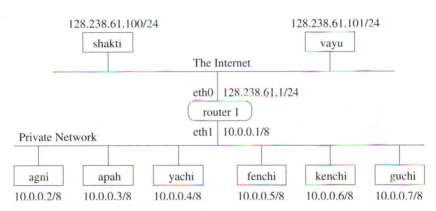

Figure 8.7. The network configuration for the NAT exercises.

add a default route in your host's routing table, using the router interface on your subnet as the default router.

One student should **telnet** to the router and configure the router as shown in Table 8.5. Note that there is a static translation that maps 10.0.0.7, or `guchi`, to 128.238.61.104.

Login to the router, execute **write term** to display the current router configuration. Execute **show ip nat translations** in the *Privileged EXEC* mode to display the translation table. Save both outputs for the lab report.

LAB REPORT   How many entries were there in the translation table? Why?

Exercise 11   Keep the login session to the router running. Execute **tcpdump -exn -nn** on all the hosts.

Before any host in the private network send any packets out, **ping** an inside host (e.g., `fenchi`) from an outside host (e.g., `vayu`). You may try to **ping** 10.0.0.5, 128.238.61.102, 128.238.61.103, or 128.238.61.104. Can you **ping** these IP addresses?

Let an inside host send packets to an outside host, e.g., from `fenchi`, execute **ping 128.238.61.100**. Can you **ping** `fenchi` from an outside host now? Why? Which IP address should be used in the **ping** command in order to **ping** `fenchi`?

Execute **show ip nat translations** in the router login window to display the translation table. Save the output for the lab report.

Exchange the data you saved with a student in the other subnet.

LAB REPORT   Answer the above questions. Use the saved translation table to justify your answers.

Compare the IP header of the ICMP query captured in the private network with that of the same ICMP query captured in the upper subnet, list their differences. Explain how NAT works.

In addition to the IP address, what else was changed in the ICMP query packet?

Exercise 12   Keep the login session to the router running. Execute **tcpdump -enx -s 100 ip proto 1 -w exc.out** to capture ICMP messages.

Execute **sock -i -u -n1 128.238.61.101 8888** on `agni` to generate an ICMP port unreachable error.

Use **ethereal** to load the `exc.out` file. Print the ICMP error message for the lab report.

Execute **show ip nat translations** in the router login window to display the translation table. Save the output for the lab report.

Exchange the data you saved with a student in the other subnet.

LAB REPORT   Analyze the IP headers, the ICMP headers, and the ICMP payloads of the ICMP port unreachable errors captured in the private network and in the public network from the first experiment. Explain how ICMP error was handled by the NAT router.

Exercise 13   Reboot the router to restore its default configuration. Then, configure the router to use PAT, as given in Table 8.6. Now all the hosts in the private network use the same IP address 128.238.61.1. However, note that there is a static translation that maps `guchi`'s port 80 to 128.238.61.1 port 80.

Table 8.6. *PAT Router Configuration in Fig. 8.7*

ip nat inside source list 8 interface ethernet 0 overload

ip nat inside source static tcp 10.0.0.7 80 128.238.61.1 80

interface ethernet 0
ip address 128.238.61.1 255.255.255.0
ip nat outside

interface ethernet 1
ip address 10.0.0.1 255.0.0.0
ip nat inside

access-list 8 deny host 10.0.0.7
access-list 8 permit 10.0.0.0 0.0.0.255

Execute **tcpdump** on all the hosts.

Generate traffic between the inside and outside hosts. Examine the **tcpdump** output to see how PAT works.

Start the Apache web server on `guchi`. Also, start the web browser `Mozilla` on an outside host (e.g., shakti), and enter the URL `http://128.238.61.1`. Save the **tcpdump** output.

Use **show ip nat translations** to display and then save the translation table.

Exchange the data you saved with a student in the other subnet.

LAB REPORT   From the **tcpdump** data, explain how PAT worked, both for a dynamic translation and a static translation.

With PAT, can you have two web servers in the private network? If not, why? If yes, explain how this can be done.

8.11 Socket programming exercises

Exercise 14   Examine the UDP socket programs `/home/guest/UDPserver.c` and `/home/guest/UDPclient.c` to learn how to write a UDP socket program.

Compile the C programs using **gcc -o UDPserver UDPserver.c -lnsl** and **gcc -o UDPclient UDPclient.c -lnsl**.

Start **tcpdump host** *remote_host* to capture packets from or to a remote host.

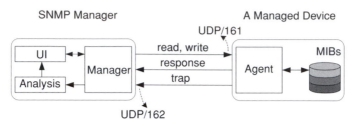

Figure 9.1. Typical SNMP operations.

Figure 9.1 illustrates a typical SNMP management scenario, consisting of an SNMP *manager* and multiple managed devices. A managed device, e.g., a host computer or a router, maintains a number of Management Information Bases (MIB), which record local management related information. An SNMP *agent* (usually a daemon on UDP port 161) runs in the managed device, providing an interface between the SNMP manager and the MIB. The SNMP manager can perform *read* or *write* operations on the elements in the MIB (or MIB *objects*) by sending SNMP messages to the agent. When the agent receives the message, it performs the required operation on the target MIB object, and returns a response to the manager. When a significant event occurs at the managed device, the agent may send a *trap* message to the manager (on UDP port 162) to report the event. An analysis tool, e.g., a graphical tool that plots the received data, may help the administrator better understand the management information collected from the managed devices.

SNMP defines several types of messages for exchanging management information between an SNMP manager and an SNMP agent. The messages are listed here.

- Get. Fetches the value of one or more objects.
- GetNext. Fetches the value of the next object after the specified object.
- Set. Sets the value of one or more objects.
- Response. Returns the value of one or more objects.
- Trap. Reports the occurrence of some significant events on a managed device. The manager does not acknowledge receptions of Traps.
- Inform. Reports the occurrence of some significant events on a managed device. The manager returns a response when an inform message is received to acknowledge it.
- GetBulk. This message allows exchanging of responses as large as possible given the constraint on message sizes. It is used to minimize the number of protocol message exchanges required to retrieve a large amount

SNMPv2 Get, GetNext, Inform, Response, Set, and Trap messages

Version Number	Community Name	PDU Type	Request ID	Error Status	Error Index	Object1	Value1	...

SNMPv2 Get Bulk Message

Version Number	Community Name	PDU Type	Request ID	Non Repeaters	Max-Repeaters	Object1	Value1	...

Figure 9.2. The SNMP message formats.

of management information. GetBulk is only available in SNMPv2 and SNMPv3.

The SNMP message formats are given in Fig. 9.2. The message fields are given here.

- `Version Number`. This is the version of SNMP. The current version is SNMPv3, but the coexistence of SNMPv1, SNMPv2, and SNMPv3 is allowed. SNMPv2 extends SNMPv1 by defining additional operations, while SNMPv3 extends SNMPv2 by adding security and remote configuration capabilities.
- `Community Name`. Defines the access scope for SNMP managers and agents. An SNMP message carrying a different community name is discarded. This provides a simple authentication for the SNMP messages.
- `Protocol Data Unit (PDU) Type`. Specifies the SNMP message type.
- `Request ID`. This field is used to match an SNMP request with the corresponding response.
- `Error Status`. This field is only set by an SNMP response. It is an integer specifying an error.
- `Error Index`. This field is only set by an SNMP response. If an error occurred, it is an integer offset specifying which object was in error.
- `Objects and Values`. A list of objects and their values.

9.2.2 The MIB structure

In SNMP, a managed device maintains a large number of SNMP objects storing management information. The Structure of Management Information (SMI) defines the rules for describing management information and the data types used in SNMP. Some examples of the data types are: (1) *Integer*,

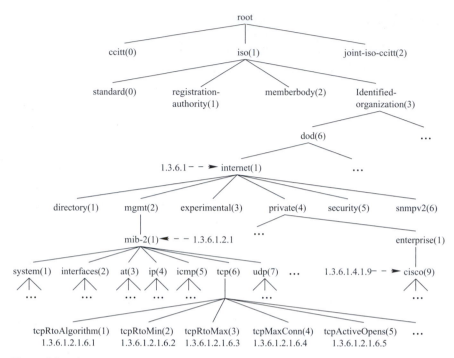

Figure 9.3. The MIB tree hierarchy.

which is a signed integer in the range of $-2,147,483,648$ to $2,147,483,647$; (2) *Octet String*, which is an ordered sequence of 0 to 65,535 octets; (3) *Sequence*, which defines a vector, with all elements having the same data type.

The objects are organized as a tree with an anonymous root, as illustrated in Fig. 9.3. Each level of the tree consists of several groups, each assigned with both a text-based name and a numerical identifier. The leaves of the `mib-2` subtree are MIB objects. Note that vendor-specific MIBs, e.g., the Cisco MIBs, are located in the *enterprise* subtree. A node (or leaf) in the tree is identified by a concatenation of the names (or IDs) of all its predecessors starting from the root. For example, the *system* node can be identified by either `1.3.6.1.2.1.1` or `iso.org.dod.internet.mgmt.mib-2.system`. The object identifiers are used in the SNMP messages to specify the target MIB objects. In Fig. 9.3, the leaf node *tcpMaxConn* is an object of Integer data type which defines the maximum number of TCP connections this system can support. It can be identified by either `1.3.6.1.2.1.6.4` or `iso.org.dod.internet.mgmt.mib-2.tcp.tcpMaxConn`.

9.2.3 NET-SNMP

NET-SNMP, formerly known as UCD-SNMP, is a very popular public domain SNMP implementation consisting of an extensible SNMP agent, a set of tools to request or set information from SNMP agents, a set of tools to generate and handle SNMP traps, and an SNMP API library for writing SNMP related programs. NET-SNMP is included in the Red Had Linux 9 installation CDs.

The SNMP MIBs are stored in the `/usr/share/snmp/mibs` directory. The SNMP agent daemon uses the `/etc/snmp/snmpd.conf` configuration file, where the community name and many other configuration options may be set. The SNMP agent daemon **snmpd** is controlled by:

/etc/init.d/snmpd start|stop.

If **snmpd** is started, MIB objects can be accessed by the following tools included in the NET-SNMP package.

- **snmpget**: Retrieves management data from a remote host, given the domain name or IP address of the remote host, the MIB object ID, and the authentication information (e.g., the community name). For example, the following command prints the time elapsed since host `shakti` was started, using the community name `test`:

  snmpget -c test -v 2c shakti.poly.edu system.sysUpTime.0.

- **snmpgetnext**: Retrieves the value of the object with the next object ID.
- **snmpset**: Sets the value of a MIB object in a remote host.
- **snmptable**: Retrieves and displays an SNMP table from a remote host.
- **snmpwalk**: Performs a series of GetNext operation, until the last object in the specified group is read.
- **snmptrap**: Sends an SNMP trap message to the SNMP manager. The traps are sent by the daemon **snmptrapd**, while the `snmptrapd.conf` file defines the events for which traps are sent.

Sun has its proprietary network management tool, called SUNNet Manager^{TM}. NET-SNMP can also run on Solaris. You can download the NET-SNMP binary package from `ftp://ftp.sunfreeware.com/` and install it, or download the NET-SNMP source code package from `http://www.net-snmp.org` and build it. Detailed instructions on installation are provided in the corresponding web sites.

9.3 Network security overview

As the Internet grows, *information security* has become a very important and challenging issue. When a computer is connected to the Internet, it is exposed to attackers from all over the world. In the Internet, many local networks are broadcast networks (e.g., Ethernet or IEEE 802.11 Wireless LAN) and Internet routers are shared by many data flows. Therefore, message exchanges between two end hosts may be intercepted or modified by an attacker. Since the Internet is a distributed network, there is no global control over all the networks and users. An attacker may claim a false identity to gain unauthorized access to information or disrupt the normal operation of a network system.

The basic network security model is shown in Fig. 9.4 where two end users communicate through an insecure network. In order to protect messages against opponents in the network, the sender may encrypt the messages using a *key* before sending them out to the network. The receiver uses the corresponding key (which could be the same key as the one used by the sender or a different key) to decrypt the message. If the keys are kept safely, the messages will not be decipherable to an opponent. In order to distribute the keys reliably, a third party which is trusted by both end users may be used. Also this third party may provide *certificates* that authenticate the users. Figure 9.5 shows the model for the network access security, where a *gatekeeper function* protects the internal information system against attacks from the outside network. In addition, the internal network performs accounting and auditing in order to detect an intrusion.

The basic elements of information security are *Authentication*, *Authorization*, and *Accounting* (AAA). Authentication is concerned with

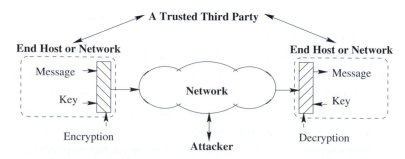

Figure 9.4. The network security model.

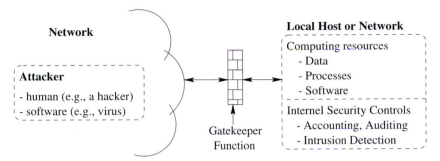

Figure 9.5. The network access security model.

ensuring that a communication is authentic, or a user is whom he or she claims to be. Authorization is concerned with assigning legitimate privilege (e.g., access to a root command or a system) to users. Accounting logs user or network behavior. The log files can then be used to detect a security intrusion. Other important security services are: *Confidentiality*, which protects transmitted data from analysis; *Integrity*, which ensures that a piece of information is not altered; *Nonrepudiation*, which ensures that the sender (receiver) cannot deny sending (or receiving) a piece of information; and *Availability*, which ensures the accessibility of a network service.

9.4 Encryption, confidentiality, and authentication

9.4.1 Data encryption

To provide the confidentiality service, a user may encrypt its messages before sending them out to the network, as illustrated in Fig. 9.4. Most classical encryption techniques use *permutation*, where the order of the plaintext characters is changed, or *substitution*, where a plaintext alphabet is mapped to a different one. The module which performs the encryption function is called a *cipher*.

Based on the operation mode, ciphers can be classified into two categories: *stream ciphers* and *block ciphers*. A stream cipher encrypts data bit by bit or byte by byte, while a block cipher first packs the data bits into a fixed length block, and then encrypts the whole block into a ciphertext block. For both categories, a key is used to encrypt the plaintext, while a corresponding key is needed to decrypt the ciphertext. Based on how the keys are used, ciphers can be classified into two categories: *symmetric-key*

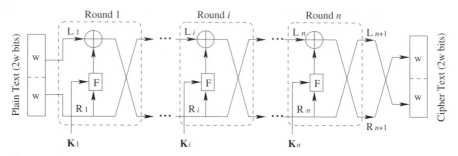

Figure 9.6. The classical Feistel network.

ciphers, where the same key is used for encryption and decryption, and *public-key* ciphers, where a *private key* is used for encryption and a *public key* is used for decryption, or vice versa. The effectiveness of the encryption schemes depends on the keys. The longer a key is, the more difficult it is to decrypt the ciphertext by brute force. Furthermore, in an open network environment with many servers and clients, any client may request service from any server. For each of the client/server connections, one or more keys may be needed. This requires keys be generated and distributed efficiently and reliably.

Most block ciphers can be characterized by the Feistel network model [13], as shown in Fig. 9.6. In this model, a $2w$ bit plaintext block is encrypted into a $2w$ bit ciphertext block. The Feistel network consists of a number of identical blocks (called *rounds*) concatenated in a chain. The plaintext is first divided into two w-bit blocks, L_1 and R_1, and fed into the first round. Each round i takes the outputs of the previous round, L_i and R_i, as inputs. The right half of the input data, R_i, is first processed with a *round function F*, using a secret key K_i. The round function F performs bit operation on the input block and the key, e.g., permutation, expansion, and exclusive-OR. Then the exclusive-OR (denoted as $\oplus$ in Fig. 9.6) of the left half of the input data, L_i, and the output of F is computed. The output of the operation is switched with R_i and fed into the next round. The same Feistel network is used to decrypt the ciphertext, with the keys applied in a reversed order, i.e., K_n is used in round 1, K_{n-1} is used in round 2, and so on so forth.

The Data Encryption Standard (DES) is the most widely used encryption standard. DES is a block-based cipher with 16 rounds, 64-bit blocks, and 56-bit keys. With DES, the 56-bit key is fed into a key generating module which computes 16 48-bit subkeys, one for each round. The strength of DES is the so called *Avalanche Effect*, where a small change in the plaintext or the 56-bit key produces a significant change in the ciphertext. This makes the ciphertext difficult to decrypt by brute force. Since the same keys are used in the encryption and decryption, DES is a symmetric cipher.

In 1998 the Electronic Frontier Foundation broke DES using a specially developed computer called the DES Cracker at a cost under $250,000. Thereafter, Triple DES (3DES) was designed to provide stronger security. 3DES uses the same DES encryption, but repeats it three times. 3DES uses three 64-bit keys, resulting in a dramatic increase in cryptographic strength. The Advanced Encryption Standard (AES) was developed to replace DES. AES uses the *Rijndael* algorithm, which is also a multiple-round, block-based cypher, but is not based on a Feistel network [14].

9.4.2 The public-key encryption schemes

Public-key encryption algorithms use two different but related keys: a private key and a public key, for each user. As their names suggest, a user's private key is kept secret, while the user's public key is distributed publicly. A ciphertext produced by a private key can only be decrypted by the public key. On the other hand, a ciphertext produced by a public key can only be decrypted by the private key. The pair of keys can be used in the following ways.

- **To provide authentication.** If user Bob wants to send a message to user Alice, he can encrypt the message using his own private key. When Alice receives this encrypted message from Bob, she can decrypt the message using Bob's public key. In this example, all other users can decrypt the message since Bob's public key is known to all. However, Alice knows that the message can only be sent by Bob, since only Bob knows his own private key.

- **To provide confidentiality.** If Bob does not want the message to be readable by other users, he can encrypt the message using Alice's public key. Alice can decrypt the message using her private key. Since no one else knows Alice's private key, the message is indecipherable to all other users.

- **To provide both authentication and confidentiality.** Bob may first encrypt the message using Alice's public key (this ensures only Alice can decrypt the message), and then further encrypt the ciphertext with his private key (this guarantees the message is from Bob). When Alice receives this message, she first decrypts the message using Bob's public key, then decrypts the results using her private key.

Table 9.1. *Key computation in RSA*

(1) Find two prime numbers p and q.

(2) $n = pq$.

(3) Find e which is less than and relatively prime to $(p - 1)(q - 1)$.

(4) $d = e^{-1} \bmod (p - 1)(q - 1)$.

The Rivest–Shamir–Adleman (RSA) scheme is a widely used, general-purpose, public-key encryption scheme. RSA uses a block cipher, where each block has a size less or equal to $B = log_2(n)$ in bits. Let M be the plaintext and C the ciphertext. Both M and C are interpreted as positive integer values during the RSA calculation. RSA encryption and decryption are performed as:

$$C = M^e \bmod n; \tag{9.1}$$

$$M = C^d \bmod n = (M^e)^d \bmod n = M^{ed} \bmod n, \tag{9.2}$$

where n is the largest binary value that a plaintext could have (i.e., $n = 2^B$). The public key consists of the pair of numbers $\{e, n\}$, while the private key is $\{d, n\}$. Any three-tuple of n, e and d that satisfies Equation (9.2) produces a pair of keys. RSA uses the scheme shown in Table 9.1 to compute the keys [13].

RSA uses the exponential function in encryption and decryption, which has higher computational cost compared with the permutation and substitution operations used in many traditional encryption schemes. In practice, public-key schemes are used in key-management and signature applications. Encrypting of the whole message using public-key encryption is not recommended.

9.4.3 Hashing and message authentication

Hashing is the operation that maps a message of variable length into a hash value with fixed length. The function defining the mapping is called a *hash function*. The difference between hashing and the schemes discussed so far is that hashing is not reversible, i.e., a hash value can be computed from a message, but the message can never be recovered from a received hash value only. The simplest hashing scheme is the block-based XOR scheme, where a message is first divided into equal-sized blocks, $[B_1, B_2, \ldots B_n]$, and then the bit-by-bit XOR operation is performed over all the blocks, i.e., $H = B_1 \oplus B_2 \oplus \cdots \oplus B_n$.

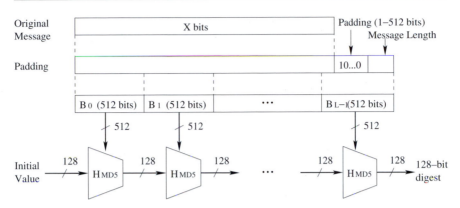

Figure 9.7. The MD5 message digest algorithm.

Many network applications require message integrity. In order to provide an integrity check and authenticate a message, hashing can be used to generate a *digest* of the message, called the Message Authentication Code (MAC). The receiver can use the digest to verify if the message is authentic.

The Message Digest 5 (MD5) algorithm is one of the most widely used hashing algorithms for providing message authentication. Figure 9.7 illustrates the MD5 algorithm. With MD5, an X-bit message is first padded so that it is just 64 bits shy of being a multiple of 512 bits long. The padding bits, which could be 1 to 512 bits, start with a "1" bit, followed by a sequence of "0" bits. Then, the 64-bit message length field follows, which is the length of the message before the padding bits were added. If the message is longer than 2^{64} bits, only the lower order 64 bits of X are used in this field. After padding, the message is divided into 512-bit blocks, each fed into a hash function. The hash functions are cascaded as shown in Fig. 9.7. The last stage produces a 128-bit message digest.

The message digest can be encrypted with the sender's private key, appended to the message, and transmitted to the receiver. The receiver, after receiving the message, may use the same MD5 algorithm to compute the digest of the received message. The original digest is decrypted using the sender's public key. If the message is genuine, the two digests should be identical.

9.4.4 Digital signatures and authentication

The above schemes assume that the two users involved in a communication are friendly. When there is a lack of complete trust between the users, *digital signatures* can be used to provide nonrepudiation service. A digital signature is analogous to handwritten signatures. It can be used to verify the date and time of a message, and to authenticate the contents of the message.

Generally, a digital signature is a bit pattern which includes a digest of the message (e.g., hashing value), the user IDs, a timestamp, and some other information. The digital signature is usually encrypted using either the symmetric encryption techniques or public-key encryption techniques. A message can be signed by the sender, and then be verified by the receiver by examining the digital signature. This approach is called the *direct digital signature* since only the communicating parties are involved. On the other hand, an arbitrator can be used to provide *certificates* to the users. This approach is called the *arbitrated digital signature*. Before a connection is established between a sender and a receiver, the sender first contacts the arbitrator (or the Authentication Server). The arbitrator validates the users and the message, and issues a *certificate* to the sender and, if required, to the receiver. The certificate may include a secret key used for this session.

The Digital Signature Standard (DSS) is a widely used digital signature scheme issued by the National Institute of Standards and Technology (NIST). It makes use of a hashing algorithm called the Secure Hash Algorithm (SHA), and the signature generated is encrypted using public-key encryption.

9.5 Application layer security

The security techniques discussed in the previous section can be deployed at different layers in the TCP/IP protocol stack. In the following, we discuss the application of such techniques in the application, transport, and network layers.

9.5.1 The Secure Shell protocol and OpenSSH

Secure SHell (SSH) is a set of protocols for secure remote login and other secure network services over an insecure network. It consists of three major components: (1) The Transport Layer Protocol [SSH-TRANS]: which provides server authentication, (2) The User Authentication Protocol [SSH-USERAUTH]: which authenticates the client-side user to the server, and (3) The Connection Protocol [SSH-CONNECT]: which multiplexes the encrypted tunnel into several logical channels. SSH will replace traditional remote access protocols, such as Rlogin, Telnet, FTP, and Remote Shell (Rsh), where security was not a major design objective. SSH supports almost any kind of public-key algorithm and various types of authentication. The SSH client and server use digital signatures to verify their identity. All communication between the client and server is encrypted.

The OpenSSH suite is a public domain implementation of the SSH protocols, including **ssh**, **scp**, and **sftp**. OpenSSH also includes the SSH daemon **sshd**, the secure FTP daemon **sftp-server**, and other basic utilities. Both Linux and Solaris platforms are supported. The OpenSSH daemon uses the `/etc/ssh/sshd_config` file for configuration, and can be turned on or off by:

/sbin/service sshd start|stop.

There are multiple OpenSSH clients, one for each remote access service. The following are the client programs.

- **ssh**: A secure client for logging into a remote machine and executing commands there. For example, to login into `shakti` as user `guest`, use:

  ssh guest@128.238.66.100.

- **scp**: A secure client for copying files between hosts. For example, to upload a file `foo.txt` to host `shakti`, use:

  scp foo.txt guest@128.238.66.100:/home/guest/foo.txt.

- **sftp**: A secure interactive file transfer program.

  OpenSSH also provides several tools for key management, including: (1) **ssh-keygen**, which is used to create keys (host keys and user keys) for public-key authentication; (2) **ssh-agent**, which is an authentication agent holding RSA keys; (3) **ssh-add**: which is used to register new keys with the SSH agent; and (4) **ssh-keyscan**: which is used to gather SSH public keys.

9.5.2 Kerberos

Kerberos is a network authentication protocol developed by the MIT Project Athena team, which uses symmetric key encryption for authenticating users for network services. Kerberos uses a trusted Authentication Server and a Ticket-Granting Server (TGS) to provide two types of tickets to a user: a *ticket-granting ticket* and one or more *service-granting tickets*, one for each network service. More specifically, Kerberos works in the following way.

1. When a user logs on to a computer, a request for the ticket-granting ticket is sent to the Authentication Server. The Authentication Server, after verifying the user ID, returns a ticket-granting ticket which is encrypted using the user's key.

2. The returned ticket-granting ticket is then decrypted using the user's key. The ticket is valid for a period of time (e.g., 10 hours), and is stored for future use. Note that the user's key is computed from the user's password. In this way, there is no need to transmit the user's password in the network.

3. When the user requests a network service, the ticket-granting ticket is used to request the corresponding service-granting ticket. The TGS uses the received ticket-granting ticket to authenticate the request, and returns the requested service-granting ticket to the user.

4. Then the user can request the network service using the service-granting ticket.

With Kerberos, the ticket-granting ticket application is performed once per user login, while the service-granting ticket application is performed once per service. In addition, the user password is not transmitted, preventing it from being sniffed by an attacker.

9.6 Transport layer and web security

As discussed in Section 8.2, the HTTP requests and responses are sent as plaintext. However, in some situations, e.g., financial transactions, extra security for the web service is needed. Web security can be provided in different ways by: (1) using the application layer security protocols, such as Kerberos; (2) using the Secure Sockets Layer (SSL) in the transport layer; and (3) using IP security (IPsec) in the Network Layer. In this section, we discuss the SSL protocol and the secure Apache server. We will discuss IPsec in the next section.

9.6.1 The Secure Sockets Layer protocol

The Secure Sockets Layer protocol (SSL) is designed to provide secure communications between a client and a server. As shown in Fig. 9.8, SSL uses TCP's reliable transport service for data communication. SSL is independent of the higher layer application protocols. Application protocols, such as HTTP, FTP, and Telnet, can use SSL for secure communication.

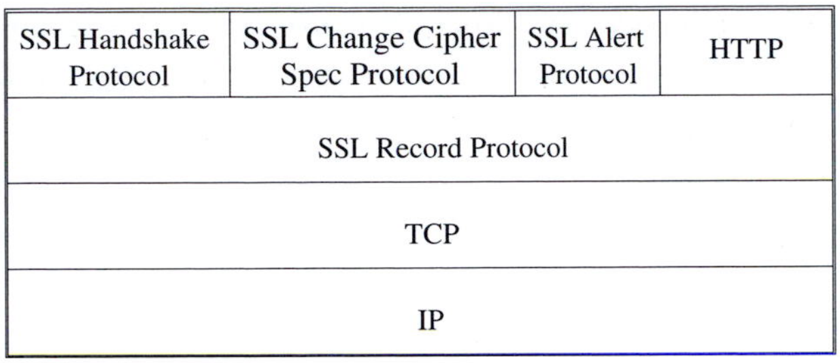

SSL Handshake Protocol	SSL Change Cipher Spec Protocol	SSL Alert Protocol	HTTP
SSL Record Protocol			
TCP			
IP			

Figure 9.8. The SSL protocol stack.

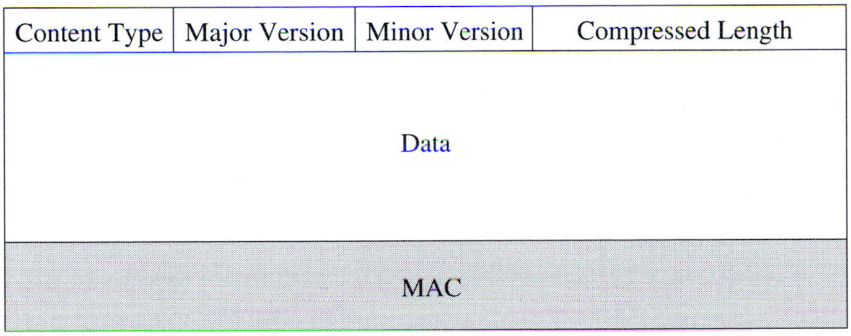

Content Type	Major Version	Minor Version	Compressed Length
Data			
MAC			

Figure 9.9. The format of a SSL record.

SSL consists of four protocols: the SSL Handshake Protocol, the SSL Change Cipher Spec Protocol, the SSL Alert Protocol, and the SSL Record Protocol, as shown in Fig. 9.8. All higher layer messages, including the messages used in the first three SSL protocols, are encapsulated in SSL *records* which are defined in the SSL Record Protocol. The SSL record header consists of an 8-bit *Content Type* field, an 8-bit *Major Version* field, an 8-bit *Minor Version* field, and a 16-bit *Compressed Length* field, as shown in Fig. 9.9. The SSL record data section consists of a Message Authentication Code (MAC) (see Section 9.4.3), the actual data, and the possible padding bytes. When a higher layer message arrives, it is first fragmented to fixed length blocks (padding may be inserted). Each block may then be compressed. The MAC is computed using the possibly compressed data, a secret key, and a 32-bit long sequence number using a hash function. Then, the data and the MAC are encrypted and the SSL record header is appended.

SSL can negotiate an encryption algorithm and session key, as well as authenticate for the secure connection. The SSL Handshake Protocol is used for the client and server to authenticate each other, to negotiate an encryption algorithm and a MAC algorithm, and to exchange the encryption keys. The SSL Change Cipher Spec Protocol is used to update the set of ciphers to be used on this connection. The SSL Alert Protocol is used to deliver SSL-related alerts to the peer entity.

9.6.2 Secure Apache server

The Apache web server can use SSL to provide a secure web service: i.e., certification of server and client, encryption of HTTP messages, etc. A secure Apache server uses TCP port 443 with URLs starting with `https:` `//`, while the unsecured Apache servers run on TCP port 80 (with the same IP address) with URLs starting with `http://`. The `mod_ssl` Apache loadable module and the **openssl** utility are needed to set up a secure Apache server. These are both preinstalled in Red Hat Linux 9.

To set up a secure Apache server, follow the steps below.

1. Create a private key: Execute the following commands:

   openssl genrsa 1024 > /etc/httpd/conf/ssl.key/server.key,
   chmod go-rwx /etc/httpd/conf/ssl.key/server.key.

   You will be prompted to create a password while running the first command. SSL uses public-key encryption for exchanging certificates and the symmetric key between the server and a client. The server encrypts its certificates using this private key and sends it to the client, along with the server's public key. The client can decrypt the certificate using the server's public key. Thus the server is certified. Next, the client generates a symmetric key, which will be used in data encryption and MAC calculations. The symmetric key is encrypted using the server's public key and sent to the server. Since only the server has the private key to decrypt this, the symmetric key is now known to the server and the client only. Next, the encryption algorithm and the hash function to use may be negotiated between the client and the server. The following data exchanges between the server and client are all encrypted using the symmetric key. Thus a secure connection is set up. These operations are defined in the SSL Handshake Protocol.

2. Create a certificate: The certificate of the server can be assigned by a certificate authority (CA), or be signed by itself. To apply a CA signed certificate, use:

   **openssl req -new -key /etc/httpd/conf/ssl.key/server.key **
   -out /etc/httpd/conf/ssl.crt/server.crt

You will be asked a number of questions regarding the server's domain name, location, organization, etc. Next, send the created certificate to a CA. The CA will verify the identity of the server and assign a new certificate for this server.

To create a self-signed certificate, go to the `/etc/httpd/conf` directory, and execute: **make testcert**.

You will be prompted for the password which was set when you created the server private key, and a number of questions regarding the server's identity. Once the certificate is created, the server can send it to a client to authenticate itself.

3. Restart the Apache server: to load the new key and the new certificate, use: **/etc/rc.d/init.d/httpd restart**

To test the secure Apache server, you can start a `Mozilla` web browser, and enter the URL `https://server_IP`. A dialog window will pop up asking if the certificate of the server should be accepted.

9.7 Network layer security

Security can be provided in the application layer, where security protocols are tailored for a specific application, e.g., SSH. On the other hand, security can also be provided in the lower layers, e.g., SSL and IP Security (IPsec), where all higher layer applications can enjoy the protection provided by a secured lower layer transparently. In the following, we discuss security support in the network layer.

IP security, or *IPsec*, is a set of protocols providing authentication and confidentiality services in the network layer. Since all protocols at the higher layers (e.g., TCP, UDP, and ICMP) have their data encapsulated in IP datagrams, IPsec protects *all* distributed applications. Higher layer protocols can enjoy the protection provided by IPsec transparently.

A typical application of IPsec is providing secure connectivity over the Internet for distributed hosts and networks. As illustrated in Fig. 9.10, two office networks (typically belonging to the same organization) can be connected by a secure channel provided by IPsec. In each office network, application data is transmitted as plaintext in regular IP datagrams. However, the data between the two office networks is encrypted and authenticated. The security-related operations, including authentication, encryption, and

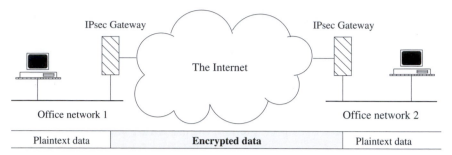

Figure 9.10. An IPsec scenario.

key management, are performed at the two IPsec-capable devices and are transparent to the users. There is no need to change or configure the hosts or the programs running in the hosts for the security service. The network shown in Fig. 9.10 is also called a *Virtual Private Network* (VPN). Compared with traditional approaches that use a leased line to connect the office networks, VPN provides good security at a lower cost.

IPsec uses two protocols to provide security: (1) an authentication protocol that uses an *Authentication Header* (AH), and (2) an encryption/authentication protocol, called the *Encapsulating Security Payload* (ESP), that encrypts the higher layer data and provides an optional authentication service. Both AH and ESP support two modes of operation: the *transport mode* and the *tunnel mode*. The Transport mode provides protection for upper-layer protocols. As shown in Fig. 9.11, the original IP header is untouched, while the remaining part of the IP datagram is either authenticated by AH, or encrypted and authenticated by ESP. The transport mode is usually used for end-to-end communication between two hosts. The tunnel mode protects the entire IP datagram. As illustrated in Fig. 9.11, a new IP header is used to route the packet, while the original IP datagram, including its header and data, are authenticated or encrypted. For the example in Fig. 9.10, a *tunnel* is established between the two IPsec gateways. An outbound IP datagram is first encrypted or authenticated, then encapsulated and forwarded in a new IP datagram. When the new IP datagram arrives at the destination network's IPsec gateway, the new header is stripped and the original IP datagram is decrypted and restored. Recall that in the MBone, similar tunnels are used to route multicast IP datagrams between two multicast islands.

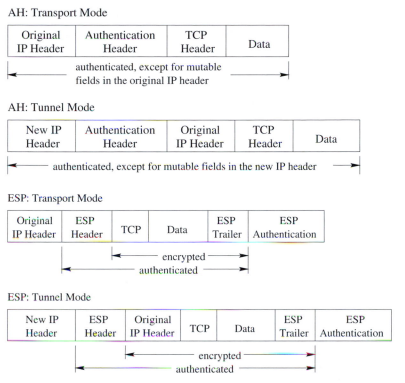

AH: Transport Mode

Original IP Header	Authentication Header	TCP Header	Data

authenticated, except for mutable fields in the original IP header

AH: Tunnel Mode

New IP Header	Authentication Header	Original IP Header	TCP Header	Data

authenticated, except for mutable fields in the new IP header

ESP: Transport Mode

Original IP Header	ESP Header	TCP	Data	ESP Trailer	ESP Authentication

encrypted
authenticated

ESP: Tunnel Mode

New IP Header	ESP Header	Original IP Header	TCP	Data	ESP Trailer	ESP Authentication

encrypted
authenticated

Figure 9.11. Encapsulation of an IP datagram using the IPsec authentication header and encapsulating security payload.

9.8 System security

9.8.1 Firewalls

A firewall is a device or program inserted between a private network and the Internet to control access. A firewall can be used to block undesired traffic from the outside, or to prevent an internal user from receiving an unauthorized external network service.

Usually the firewall is the only access point of a private network, i.e., all outbound traffic from the private network should be routed through the firewall in order to enforce the desired access control. There are three types of firewalls. A *packet filter* blocks selected network packets. An *application gateway*, or a *proxy server*, is mainly used to regulate outbound traffic. A proxy server acts as a relay for a specific application (e.g., web service). The traffic between the internal client and the remote server (e.g., HTTP requests and responses) is relayed by the proxy server. A *circuit-level gateway* acts like a switch board, switching an internal connection to another external connection.

9.8.2 iptables

Packet filtering is built into the Linux kernel. The default firewall in Linux kernel 2.4 is the **iptables** (also called the *netfilter*). A firewall policy, also called a *rule*, consists of two components: a condition (e.g., destination port number of a packet) and the operation (called the *target*) on the packets that satisfy the condition (e.g., drop). In Linux, rules are organized into three *tables*, based on the operations to be performed. The *filter* table is the default table for filtering packets. The *nat* table is used to alter packets that create a new connection, while the *mangle* table is used for some specific types of packet alteration. In each table, rules are further organized into several *chains*, based on the type of packets they regulate. For example, the filter table has three chains: the INPUT chain which consists of the rules applied to the packets destined to the host, the OUTPUT chain which consists of the rules applied to the packets generated by the host, and the FORWARD chain which consists of the rules applied to the packets routed by the host (when the host is configured to be a router).

In **iptables**, a packet is first dispatched to the corresponding chain. Then the packet is checked against each rule in that chain, one at a time. If there is a match, the target defined in that rule is performed on that packet. Rules in the tables are configured by running the **iptables** command, which defines the packet type and the corresponding target. The syntax of **iptables** is:

iptables *[-t table-name] command chain-name parameter1 option-1 . . . *
parameterN optionN

The *table-name* could be filter, nat, or mangle. Commands indicate what kind of application to perform on the table, e.g., **-A** means appending the rule to the end of the specified chain and **-F** flushes the selected chain (i.e., deletes all the rules in that chain). The next parameter, *chain-name* is the name of the chain to configure, i.e., INPUT, OUTPUT, or FORWARD in the filter table. Next is a list of parameters and options, which defines the rule. Several frequently used parameters are listed.

- **-d**: indicates the destination hostname, IP address or network address of a packet.

- **-s**: indicates the source hostname, IP address or network address of a packet.
- **-i**: indicates the incoming interface of a packet.
- **-p**: indicates the IP protocol for the rule, which could be `icmp`, `tcp`, `udp` or `all`.

The target option could be the following.

- **ACCEPT.** Allows the packet to successfully move on to its destination or another chain.
- **DROP.** Drops the packet without notifying the sender.
- **QUEUE.** Queues the packet to be handled by a user-space application.
- **RETURN.** Stops checking the packet against the rules in the current chain.
- **LOG.** Logs all packets that satisfy this rule.
- **REJECT.** Drops the packet and notifies the sender.

The command **iptables -h** displays a detailed list of the parameters and options.

After configuring the tables, you can restart **iptables** to load the new rules by: **/sbin/service iptables restart**. Or you can save the rules by **/sbin/service iptables save**. The rules will be saved in the `/etc/sysconfig/iptables` file. The **/sbin/chkconfig –level 345 iptables on** command makes the system start **iptable** when it bootstraps.

9.8.3 Auditing and intrusion detection

There are many ways to check if a system is compromised. For example, all Unix and Linux systems log network events and user activity. By examining the log files, an intruder may be identified. Usually the log files are stored in the `/var/log` directory. The system logging daemon is **syslogd**, which supports both local and remote logging. Red Hat Linux 9 provides a graphical interface tool, **redhat-logviewer**, for browsing the system logs.

Both Red Hat Linux and Solaris support the following commands to monitor active users or check network services.

- **users**: displays a list of user names currently logged on.
- **who**: displays information about currently logged-on users.
- **last**: displays data about current and recent logins.
- **netstat -l**: lists listening sockets.

- **chkconfig ——list**:[1] lists services named in the `/etc/init.d/` directory and their status.

A public domain tool called `Tripwire` can provide an integrity check on the system files. If a system file is changed by an attacker or a virus, `Tripwire` can detect and report the change. `Tripwire` is available from `http://tripwire.org`.

9.9 SNMP exercises

For the exercises in this section, the network topology is given in Fig. 1.3, where all the hosts are connected in a single network segment using their default IP addresses, i.e., from 128.238.66.100 to 128.238.66.107.

Before the lab, the lab instructor should:

1. Backup the original **snmpd** configuration file:

      mv /etc/snmp/snmpd.conf /etc/snmp/snmpd.conf.save.

2. Create a simple configuration file `/etc/snmp/snmpd.conf` with a single line defining a read-only community guest:

```
rocommunity guest
```

Exercise 1 Use **pgrep snmpd** to check if **snmpd** is started. Try to stop and then start the SNMP agent daemon using **/etc/init.d/snmpd start|stop**.

Study the **snmpd** configuration file `/etc/snmp/snmpd.conf`. Also study the default configuration file `/etc/snmp/snmpd.conf.save`. This file is well commented. Read the comments and study the configuration options.

Study the MIB files in the `/usr/share/snmp/mibs` directory. Examine the Interface MIB `IF-MIB.txt` and the TCP MIB `TCP-MIB.txt` to see the MIB objects and data types. Save these two files for the lab report.

LAB REPORT   What is the community name used in this lab? What is the use of the community name?

What is the data type for the MIB object `ifMtu.2`? What is the definition of the MIB object `ifPhysAddress` and `ifInOctets`?

What is the data type and definition of `tcpRtoAlgorithm`? What values are allowed for `tcpRtoAlgorithm`? What is the definition of `tcpMaxConn`?

---

[1] Not available in Solaris.

Exercise 2   Use **snmpwalk -v 2c -c guest localhost interface** to display the Interface MIB.

Use **snmpwalk -v 2c -c guest localhost tcp** to display the TCP MIB.

You may run **man snmpwalk** to find out the meanings of the options used in the commands. Compare the outputs with the MIB files you saved in the previous exercise. Also compare the outputs of the first command with that of **ifconfig -a**.

Retry the **snmpwalk** commands, but change `guest` to `public`. Can you display the MIBs this time?

LAB REPORT   What is the MTU of the Ethernet interface? What is the MTU of the loopback interface? Justify your answer with the **snmpwalk** output and the **netstat** output.

Why did the **snmpwalk** command with a community name `public` fail?

Exercise 3   Execute **tcpdump udp port 161 -w ex3.out** to capture SNMP messages.

Execute **snmpget -v 2c -c guest** *remote_host* **IF-MIB::ifMtu.1** to get the MIB object `IF-MIB::ifMTU.1` from a remote machine.

Save the **snmpget** output and terminate **tcpdump**.

Use **ethereal** to load the `ex3.out` file and analyze the format of the captured SNMP Get and Response messages. Print the messages for the lab report.

LAB REPORT   What is the port number used by the SNMP agent?

What are the full text-based and numerical object ID's of the MIB object `interface.ifMTU.2`? What was the value returned? Justify the answer using Fig. 9.3 and the **ifconfig** output.

Draw the format of one of the SNMP messages saved, including the name and value of each field.

9.10 Exercises on secure applications

Exercise 4   Execute **tcpdump -enx -s 100 -w ex4a.out host** *your_host* **and** *remote_host* to capture packets between your machine and a remote machine.

Execute **ftp** *remote_machine*. When prompted, type "1111" for the login ID, and "2222" for the login password. Then terminate **tcpdump** and **ftp**.

Use **ethereal** to load the `ex4a.out` file. Analyze and print the packets that carry the login ID and the password for the lab report.

Repeat the above experiment, but use **telnet** and save the **tcpdump** output in the `ex4b.out` file.

LAB REPORT   Can you see the login ID and the password in the FTP experiment? Submit the two packets you printed.

Can you see the login ID and the password in the TELNET experiment? Submit the packets you printed.

What is the difference between FTP and TELNET in their transmission of user ID's and passwords? Which one is more secure?

Exercise 5   Execute **tcpdump -enx -s 100 -w ex5a.out host** *your_host* **and** *remote_host* to capture packets between your machine and a remote machine.

Execute **sftp** *remote_machine*. When prompted, type "yes" to continue the connection and "1111" for the login password. Then terminate **tcpdump**.

Use **ethereal** to load the ex5a.out file. Analyze and print one or two SSH packets for the lab report.

Repeat the above experiment, but use **ssh** and save the **tcpdump** output in the ex5b.out file.

LAB REPORT   In each experiment, can you extract the password from the **tcpdump** output? Can you read the IP, TCP, SSH headers? Can you read the TCP data?

What is the client protocol (and version) used in both cases?

What is the port number used by the **ssh** server? What is the port number used by the **sftp** server? Justify your answer using the **tcpdump** output and the /etc/services file.

9.11 Exercises on secure Apache server

In the exercises in this section, teams of two students work together using two workstations.

Exercise 6   Run **man openssl** to study the OpenSSL command line tool.

Create a new private key for the Apache server, using:

openssl genrsa 1024 > /etc/httpd/conf/ssl.key/server.key.

To create a self-signed certificate, go to the /etc/httpd/conf directory, and execute: **make testcert**.

Then you will be asked a number of questions, regarding the location, affiliation, etc. of the Apache server. After you type in the answers, a self-signed certificate is created at /etc/httpd/conf/ssl.crt/server.crt.

Save the **make** output for the lab report.

Exercise 7   Restart the Apache server to load the new key and the new certification: **/etc/rc.d/init.d/httpd restart**.

Execute **tcpdump -w https.out host** *your_host* **and** *remote_host* to capture the packets between your host and a remote host.

On the remote host, start the `Mozilla` web browser. After typing in the URL `https://your_host`, a dialog window titled "Website Certified by an Unknown Authority" will pop up, reporting the reception of a certificate signed by an unknown authority and asking if you want to continue.

Click the "View Certificate" button. Then a "Certificate Viewer" window pops up, displaying detailed information about the received certificate. Examine the certificate and dump the window into a picture if necessary (see Section 1.3.3 on how to dump a window). Save the pictures for the lab report.

Click the "Continue" button in the "Website Certified by an Unknown Authority" dialog window to accept the certificate. Then terminate **tcpdump** and `Mozilla`.

Use **ethereal** to load the `https.out` file and examine the operation of SSL.

LAB REPORT   What is the port number used by the secure Apache server?

Compare the general information of the received certificate with the **make** output saved in the last exercise. Are they consistent?

What is the `Subject` of the received certificate? Who is the `Issuer` of this certificate? Are they the same?

What is the `Certificate Signature Algorithm` used to generate and distribute this certificate?

When was the certificate signed? When will it expire?

9.12 Exercises on Firewalls and Iptables

In this exercise, students pair up to work together using two workstations.

Exercise 8   Execute **iptables -L -v** to list the existing rules in the filter table. Save the output for the lab report.

Append a rule to the end of the INPUT chain, by executing

iptables -A INPUT -v -p TCP ――dport 23 -j DROP.

Run **iptables -L -v** again to display the filter table. Save the output.

On both machines in your group, execute **tcpdump host** *your_host* **and** *remote_host*. Then, **telnet** to the host where the rule is set from the remote machine. Save the **tcpdump** output for the lab report.

LAB REPORT   Can you **telnet** to the host from the remote machine?

From the **tcpdump** output, how many retries did **telnet** make? Explain the exponential backoff algorithm of TCP timeout and retransmission.

Exercise 9 Delete the rule created in the last exercise, by:

iptables -D INPUT -v -p TCP –dport 23 -j DROP.

Then, append a new rule to the INPUT chain:

**iptables -A INPUT -v -p TCP ––dport 23 -j REJECT **
––reject-with tcp-reset.

Execute **iptables -L -v** to display the new rule.

On both machines in your group, restart **tcpdump**, and then **telnet** to the host where the rule is set from the remote machine. Save the **tcpdump** output for the lab report.

LAB REPORT   Explain the difference between the **tcpdump** outputs of this exercise and the previous exercise. How many attempts did TCP make this time?

9.13 Exercises on auditing and intrusion detection

Exercise 10 Start the graphical interface tool **redhat-logviewer** to examine the log files in your host. If a log (e.g., the Apache Access Log) is too long, type a keyword (e.g., GET) in the "Filter for" field to display those log entries containing the keyword. Enter the keyword "failed" to display logged failures.

Go to menu Edit/Preferences . . . to see where the log files are stored.

Exercise 11 Red Hat Linux uses a utility called Webalizer to analyze the web server log files. Webalizer reads the Apache log files and creates a set of web reports on server statistics. It is pre-installed in Red Hat Linux 9.

To view the reports, start Mozilla and enter the URL http://localhost/usage/ index.html. Examine the web statistics displayed in the browser. Also click on the *month* links in the Summary by Month table to see the statistics of each month.

Next, enter the URL http://*remote_host* /usage/index.html to view the reports on the remote machine.

LAB REPORT   List the most frequently visited pages at the local Apache server and the remote Apache server during the most recent month, respectively.

List the web pages that have the most number of bytes transferred by the local and the remote server during the most recent month, respectively.

Exercise 12   Execute **netstat -l** to display the listening sockets in your host.

Execute **chkconfig ――list** to list the services in the `/etc/init.d/` directory and their status. Save the output for the lab report.

LAB REPORT   Is the **rlogin** service enabled in your host?

References and Further reading

References

1. V. C. Cerf and R. E. Kahn, A protocol for packet network interconnections, *IEEE Transactions on Communications*, **COM-22**:5, (1994) 637–48.
2. J. Reynolds and J. Postel, *The Request for Comments Reference Guide*, IETF Request For Comments 1000, August 1987. [Online]. Available at: http://www.ietf.org.
3. C. Huitema, *IPv6: The New Internet Protocol*. Available at: http://csrc.nist.gov/publications/fips/fips197/fips-197.pdf. (Prentice Hall, 1998).
4. G. R. Wright and W. R. Stevens, *TCP/IP Illustrated, Volume 2: The Implementation*. (Reading, MA, USA: Addison-Wesley, 1995).
5. W. R. Stevens, *TCP/IP Illustrated, Volume 1: The Protocols*. (Reading, MA, USA: Addison-Wesley, 1994).
6. Cisco IOS documentation, *Cisco IOS Configuration Fundamentals Command Reference – Release 12.2*. [Online]. Available at: http://www.cisco.com.
7. Cisco IOS documentation, *Cisco IOS Configuration Fundamentals Configuration Guide – Release 12.2*. [Online]. Available at: http://www.cisco.com.
8. V. Paxson and M. Allman, *Computing TCP's Retransmission Timer*, IETF Request For Comments 2988, November 2000. [Online]. Available at: http://www.ietf.org.
9. D. Chiu and R. Jain, Analysis of the increase and decrease algorithms for congestion avoidance in computer networks, *Computer Networks and ISDN Systems*, **17**, (1989), 1–14.
10. H. Schulzrinne, A. Rao and R. Lanphier, *Real Time Streaming Protocol*, IETF RFC 2326, April 1998. [Online]. Available at: http://www.ietf.org.
11. J. Rosenberg, *et al.*, *SIP: Session Initiation Protocol*, IETF RFC 3261, June 2002. [Online]. Available at: http://www.ietf.org.
12. R. Stevens, *UNIX Network Programming, Volume 1: Network APIs: Sockets and XTI*, 2nd edn. (Upper Saddle River, NJ, USA: Prentice Hall, 1998).
13. W. Stallings, *Cryptography and Network Security: Principles and Practice*, 2nd edn. (Upper Saddle River, NJ, USA: Prentice Hall, 1999).

14. National Institute of Standards and Technology, *Announcing the Advanced Encryption Standard (AES)*, Federal Information Processing Standards Publication 197, November 2001.

Further reading

15. The IETF website. [Online]. Available at: `http://www.ietf.org`.

16. The Linux Documentation Project website. [Online]. Available at: `http://www.tldp.org/`.

17. The Cisco documentation website. [Online]. Available at: `http://www.cisco.com/univercd/home/home.htm`.

18. Sun, *Solaris 8 System Administration Guide, Volume 1, Volume 2, and Volume 3*. [Online]. Available at: `http://www.sun.com`.

19. Get IEEE 802 website. [Online]. Available at: `http://standards.ieee.org/getieee802/`.

20. Cisco IOS documentation, *Cisco IOS Bridging and IBM Networking Configuration Guide*. [Online]. Available at: `http://www.cisco.com`.

21. The NIST Role-Based Access Control website. [Online]. Available at: `http://csrc.nist.gov/rbac/`.

22. J. H. Saltzer, D. P. Reed and D. D. Clark, End-to-end arguments in system design. [Online]. Available at: `http://www.reed.com/Papers/EndtoEnd.htm`.

Appendix A: instructor's guide

Finally, after years of working with network programming, I came to realize that 80% of all network programming problems were not programming problems at all, but were from a lack of understanding of how the protocols operate.

I also realized that (there) were numerous publicly-available tools out there that aid in understanding the protocols and anyone could use them, when shown how.

W. Richard Stevens

A.1 Lab operation mechanism

This guide is based on a lab course offered at Polytechnic University, Brooklyn, New York, USA. The course personnel were a faculty member who delivered lectures on TCP/IP protocols and two lab instructors who were responsible for the setup and maintenance of the lab, and assisting students during experiments. Each experiment, consisting of the exercises in one chapter of this book, typically took the students two to three hours to complete.

Due to the limitation of space and equipment, the whole class was divided into small groups of eight (i.e., the number of workstations in the lab). The groups were scheduled to use the lab in different time slots. In the labs, each student was assigned a workstation and was required to perform the experiments independently. If required, two or more students could share a workstation and perform the experiments together.

Students were required to read the corresponding chapter in this book before the lab experiments. Pre-lab reading and preparation result in the students getting the most from the exercises, rather than blindly following the instructions without understanding the underlying concepts. For each experiment, they did the following.

1. Interconnected hosts using hubs, bridges, routers, and cables/connectors to build networks with various topologies.
2. Configured the hosts, the bridges, and the routers, e.g., setting the host or router IP addresses, choosing which protocol to run, etc.
3. Ran network applications to generate network traffic related to the protocol being studied.
4. Ran diagnostic applications simultaneously to monitor and capture packets in the network, and saved the collected data.
5. Analyzed data and wrote a lab report after the experiments.

Throughout the laboratory sessions, students were required to carry out the following.

1. Bring a textbook if being used, this book, and a 3.5″ floppy disk to each session of the laboratory. The floppy disk may be used to copy experimental data since the hosts in the lab may not be connected to the Internet.
2. In most of the exercises, use the *dotted-decimal* IP addresses rather than the machine names (e.g., use 128.238.66.100, instead of shakti), since we did not run a domain name server, and did not want to change the /etc/hosts file for each experiment.
3. The laboratory report should be word-processed. Experimental data may not be copied by hand, although minor handwritten corrections were allowed. In addition, the laboratory report should have the name of the workstation assigned to the student and the date the experiment was performed.
4. Do not include unnecessary data that has no direct bearing on the results reported in the report. Submit the output data only if it is related to the answer.
5. Do not turn off the workstation. If workstations are not shutdown properly, the file system may be damaged. If there is a problem with a workstation or a router, contact the lab instructor.

A.2 Lab equipment

The Internet consists of host computers, hubs, switches, bridges, and routers. In the experiments, we build various networks using these devices. The equipment used in the lab included the following items.

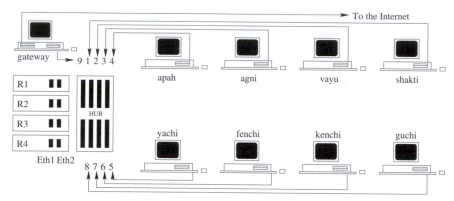

Figure A.1. An overview of the TCP/IP networking lab equipment.

1. Eight[1] desktop PCs with the Red Hat Linux 9.0 (or a later version) operating system, each with an Ethernet interface card.

You can also set up the TCP/IP lab with eight Workstations with the Solaris 8 (or a later version) operating system, each equipped with an Ethernet interface card.

---

2. Four[1] Cisco 2600 series routers with Cisco IOS release 12.0. Each router has two Ethernet interfaces and a console port for local configurations. These routers will be configured to work as bridges in Chapter 3, and as routers in Chapters 2, 4, and 5, respectively.
3. Eight Ethernet Hubs. These hubs are used to build Ethernet LAN segments with the star topology. Each hub should have at least four ports. In some exercises, more ports are required to connect all the hosts (and some router ports) to form a single LAN segment. In these cases, two or more hubs could be concatenated to extend the number of ports.
4. A number of Ethernet Cables. We use 10 Base-T cables with RJ-45 connectors. In addition, a RJ-45-to-RJ-45 *crossover* cable and a RJ-45-to-DB-9 adapter are needed in order to configure a router through the *console* port. These generally come with a new router.

Figure. A.1 is an overview of the lab equipment. As shown in the figure, we name the computers and the routers for easy exposition. The routers and hubs are installed on a rack at one end of the room. The computer

---

[1] We used eight workstations in our lab. However, a minimum setup consisting of five workstations and two routers is needed (see Section A.4).

Table A.1. *Host IP and MAC addresses*

Label	Host Name	IP Address	MAC Address
1	shakti	128.238.66.100	
2	vayu	128.238.66.101	
3	agni	128.238.66.102	
4	apah	128.238.66.103	
5	yachi	128.238.66.104	
6	fenchi	128.238.66.105	
7	kenchi	128.238.66.106	
8	guchi	128.238.66.107	

named `gateway` is used for Internet access, which is optional and will be discussed in section A.6.

This guide focuses on the Internet protocols rather than specific products. In the experiments, we try to avoid the proprietary aspect of the hardware and the operating system used. With minimal modifications, this guide can be used with other systems, e.g., PCs with the Mandrake Linux or the FreeBSD operating system.

A.3 Software installation and configuration

A.3.1 Operating system

To install Red Hat Linux 9.0 on a desktop computer, you need to insert the Red Hat Linux 9.0 CD into the CDROM Driver, and boot the computer from the CDROM. Then, follow the instructions from the installation program. In most cases, we use the default configurations. Therefore you just need to click the `Next` button to continue the installation process. The following is a list of the lab-specific selections.

1. You need to give a host its domain name and an IP address for the Ethernet interface. You may refer to Fig. A.1 and Table A.1 for the host names and IP addresses used in this guide.
2. Do not enable DHCP.[2] Do not install DNS. Do not give a `gateway` IP address.
3. Disable the firewall.
4. Choose the *server* installation, including the *kernel source*.

---

[2] But you need to install both the DHCP server and the client.

5. After the installation, create a user named `guest`, with a password of `guest1`. Choose `bash` as the login shell for user `guest`. See Section A.5 for how to assign partial root privilege to user `guest`.

   Edit the `.bash_profile` file, change the line that defines the `PATH` environment variable to:

   PATH=$PATH:$HOME/bin:/usr/sbin:/sbin:/usr/local/bin .

   Now user `guest` can access the diagnostic tools without typing the full path. Note that you need to run **source .bash_profile** for the new `PATH` to take effect. You can verify the new `PATH` by: **echo $PATH**. When you install a new program, you may also append the directory of the executable to the `PATH` variable.

6. Enable the following services, using the **chkconfig** command as root:[3]

   Ftp: **/sbin/chkconfig vsftpd on,**

   Telnet: **/sbin/chkconfig telnet on,**

   Echo: **/sbin/chkconfig echo on,**

             /sbin/chkconfig echo-udp on,

   Time: **/sbin/chkconfig time on,**

             /sbin/chkconfig time-udp on.

7. If the Linux system installed does not support multicast by default, you may need to recompile the kernel. Make sure the `IP: multicasting` option is enabled before the kernel is compiled.

Installing Solaris 8.0 is similar to the above. Basically, you can use all the default settings for each installation step. When the installation finishes, you need to create a new user named `guest` with a password of `guest1`, and set the `PATH` environment for `guest`.

To enable the above services in Solaris, you need to edit the `/etc/inetd.conf` file. Make sure that the lines corresponding to these services are uncommented. Then reboot the machine, or run **pkill -HUP inetd** to let **inetd** reload the new configuration.

Initially the workstations are connected in a single LAN segment, as shown in Fig. 1.3. The IP addresses of the host interfaces are given in Table A.1. You need to insert the MAC addresses of the host interfaces.

---

[3] You can enable these services using the graphical configuration tool as well. Invoke the system menu by clicking the Red Hat logo at the lower left corner of the workspace. Then go to /System Settings/Server Settings/Services as root. A dialog pops up with a list of services available from this host. You need to simply select the services you want to enable, e.g., Telnet, then reboot the computer.

A.3.2 Diagnostic tools

Many TCP/IP implementations, e.g., Red Hat Linux and Sun Solaris, provide some useful tools for network maintenance and diagnosis. In addition to these tools, we use several other tools in the experiments. All the additional tools used in the lab are in the public domain, and can be downloaded for free.

The following is a list of the diagnostic tools used in the lab, as well as the links to download them and a brief installation guide. More information on installation and usage can be found at the websites given below.

sock

Sock is a network traffic generator written by W. Richard Stevens. It can be downloaded from W. Richard Stevens' homepage: `ftp://www.kohala.com/start/unpv12e.html`. You can download the source code package and build the **sock** program locally. The README file coming with the package gives detailed instructions on building **sock**. Basically, you need to build the libraries first, and then build **sock**.

The *Transaction TCP (T/TCP) for Linux* project website provides a refined version of **sock** at: `http://ttcplinux.sourceforge.net/tools/tools.html`. Both the binary executable and the source code package are available. In fact, you can simply download the binary code and use it in the lab.

tcpdump

Tcpdump is a command-line-based network traffic sniffer and analyzer. It is preinstalled in Red Hat Linux 9.0 at `/usr/sbin/tcpdump`.

If you are using Solaris, or your Linux system does not have a preinstalled **tcpdump**, you need to install the program. **tcpdump** depends on the **libcap** package which captures packets from a network interface. You can download both the rpm packages or the source code packages for **libcap** and **tcpdump** from the **tcpdump** homepage: `http://www.tcpdump.org`. If you download the source code package, you need to build **tcpdump** locally, by running **./configure**, **make**, and **make install**.

ethereal

Ethereal is a network traffic sniffer and analyzer with a graphical user interface. It has similar functions as **tcpdump**, but with an enhanced user interface. Ethereal depends on two packages: the GTK+ version 1.2 or later for the graphical user interface, and the libcap package for packet capture. Both packages are preinstalled in Red Hat Linux 9.0.

Ethereal is included in the Red Hat Linux 9 installation CDs. To install, insert the Red Hat Linux Installation CD 1 into the CD drive and reboot the system. In the following Anaconda Red Hat Linux installer, choose Perform an upgrade of an existing installation, customize packages to be upgraded, and check Skip boot loader updating. In the following Individual Package Selection window, choose **ethereal** and click the Next button. Then Ethereal will be installed.

Alternatively, you can download the source package, ethereal-0.9.10.tar.gz (or a later version), from the Ethereal website: http://www.ethereal.com. Then,

1. Run **gunzip ethereal-0.9.10.tar.gz** and **tar -xvf ethereal-0.9.10.tar** to uncompress and extract the tar ball.
2. Change to the ethereal-0.9.10 directory, run **./configure**, **make**, and **make install**.

The executable is installed at /usr/local/bin/ethereal. A companion command-line tool, **tethereal**, is installed in the same directory. **tethereal** is useful when the graphical environment, i.e., gnome or XWindows, is not available.

Glib and GTK+ packages are also included in the Solaris Companion Software CD, or are downloadable from http://www.sun.com/software/solaris/freeware/. You need to make sure that these two packages are installed. Then, uncompress and install the package by **bunzip ethereal-x.y.z-solaris2.9-sparc-local.bz2** and **pkgadd -d ethereal-x.y.z-solaris2.9-sparc-local**.

The routed Routing Daemon

The **routed** rpm package is included in the Red Hat Linux 9 installation CDs. You can install the package from the CDs, or download it from, e.g., http://rpmfind.net/linux/RPM/redhat/severn/i386/routed-0.17-15.i386.html and install it using: **rpm -ivh routed-0.17-15.i386.rpm**. The program is installed at /usr/sbin/routed. If you wish to start **routed** every time when the system boostraps, execute **chkconfig routed on** as root.

In Solaris, the RIP routing daemon is `/usr/sbin/in.routed`, which is preinstalled.

The TFTP client and server

TFTP is used in Chapter 5. To install TFTP,

1. Download the TFTP client package (e.g., `tftp-0.32-4.i386.rpm`) and the server package (e.g., `tftp-server-0.32-3.i386.rpm`) from the Red Hat website: `http://www.redhat.com/swr/`. Note that the rpm packages are also included in the Red Hat Linux 9 installation CDs.
2. Run the following as root: **rpm -ivh tftp-0.32-4.i386.rpm** and **rpm -ivh tftp-server-0.32-4.i386.rpm**.

Now the TFTP client and server are installed. A directory named `/tftpboot` is created for files that you want to distribute using TFTP. Then, change the line `disable=yes` to `disable=no` in the TFTP configuration file `/etc/xinetd.d/tftp`. Finally, you need to restart the xinetd daemon, by running the following command as root: **/etc/rc.d/init.d/xinetd restart**.

TFTP is preinstalled in Solaris 8.0. Generally it is disabled due to security concerns. The following instructions show how to enable the TFTP server.

1. Turn on the `in.tftpd` daemon by creating the `/tftpboot` directory, i.e., run the following as root: **mkdir /tftpboot**.
2. Create a symbolic link to the directory, by: **ln -s /tftpboot/. /tftpboot/ tftpboot**.
3. Uncomment the TFTP line in the `/etc/inetd.conf` file. This line should look like the following:

```
tftp dgram udp wait root /usr/sbin/in.tftpd in.tftpd -s /tftpboot
```

Reboot the host, then a tftp client can only download files in the `/tftpboot` directory.

JMStudio

JMStudio is a Java-based realtime streaming tool from Sun Microsystems. It is provided as a demonstration of the Sun Java Media Framework (JMF). We use JMStudio in the multimedia multicasting exercises in Chapter 7.

Installing Java Development Kit (JDK)

Since JMStudio is a Java program, you need to have JDK for Linux (or a later version) installed first. You can download the Linux version of Java 2 Standard Edition (J2SE) 1.4.2 from `http://java.sun.com/j2se/1.4.2/download.html`. Then, get into the directory where this installer file is stored, and run the following to make it executable:

chmod 755 j2sdk-1_4_2-nb-3_5_1-bin-linux.bin.

Next, type the following command to start the installation process:

./j2sdk-1_4_2-nb-3_5_1-bin-linux.bin

You will be asked several questions by the installation wizard. You must agree to the license agreement, and may choose the default answer in all other cases. When the installation is over, you need to append the directory of JDK executables, e.g., `/opt/j2sdk_nb/j2sdk1.4.2/bin`, to guest's `PATH` environment variable.

Installing JMF

JMF works with Red Hat Linux 6.2 or a later version. The Linux JMF package, (`jmf-2_1_1e-linux-i586.bin`), can be downloaded from `http://java.sun.com/products/java-media/jmf/2.1.1/download.html`. Then, get into the directory where this installer file is stored, and run the following:

chmod +x ./jmf-2_1_1e-linux-i586.bin.

Next, run the installer to extract JMF to a directory, e.g., `/home/LAB/`. A directory called `JMF-2.1.1e` will be created in this directory. Then in the configuration file `/home/guest/.bash_profile`, set the CLASSPATH environment parameter to reference the JMF directory:

```
JMFHOME=/home/LAB/JMF-2.1.1e
export JMFHOME
CLASSPATH=$JMFHOME/lib/jmf.jar:.:$CLASSPATH
export CLASSPATH
```

Next, set the `LD_LIBRARY_PATH` environment parameter to reference the JMF libraries:

```
LD_LIBRARY_PATH=$JMFHOME/lib:$LD_LIBRARY_PATH
export LD_LIBRARY_PATH
```

Now JMF is installed. You can find the JMFStudio executable, **jmstudio**, in the `$JMFHOME/bin/` directory.

The DBS TCP Benchmark

Distributed Benchmark System (DBS) is a TCP performance measurement tool. We use DBS as a TCP traffic generator in Chapter 6.

Package Dependencies

DBS depends on three software packages. It uses `perl` 5.0 or a later version and `gnuplot` to plot the measured traces. These two packages are preinstalled in Red Hat Linux 9.0. DBS also uses `ntp` to synchronize the clocks of all the participating hosts, which is available from `http://www.ntp.org/downloads.html`. Similarly, `ntp` is preinstalled in Red Hat Linux 9.0 as well.

Both `perl` and `ntp` are preinstalled in Solaris 8.0. `gnuplot` is included in the Solaris Companion CD free software and can be downloaded from `http://www.sun.com/software/solaris/freeware/pkgs_download.html`.

Installing DBS

You can download the source code package of DBS from `http://www.kusa.ac.jp/~yukio-m/dbs/download.html`. DBS supports Linux kernel 2.0.* or later, as well as Sun OS 4.1.3, 4.1.4, and 5.5.*, and FreeBSD. To install DBS:

1. After you download the `dbs-1.2.0beta1.tar.gz` file, extract it by: **gun-zip dbs-1.2.0beta1.tar.gz** and **tar -xvf dbs-1.2.0beta1.tar**. A directory called `dbs-1.2.0beta1` is created, which contains all the DBS files.
2. Go to directory `../dbs-1.2.0beta1/src`, and run **make**.
3. Run **make install** as root to copy the executables to the `/usr/local/etc` directory. You may add the `/usr/local/etc` to user guest's PATH environment.

When the installation is over, you need to delete the first two lines in the `/usr/local/etc/dbs_view` file. The first line is "#!/usr/local/bin/perl", and the second line is "#!/usr/local/bin/perl -d". These two lines point to a wrong directory for **perl**.

When installing DBS on Solaris, you need to uncomment the LDFLAGS line in the Makefile. The first two lines in the /usr/local/etc/dbs_view file should not be deleted.

The NIST Net emulator

NIST Net is a Linux-based network emulation package. NIST Net sets up a single Linux computer as a router and perform firewall-like functions, to emulate a wide variety of network conditions. We will use NIST NET to perturb the normal operations of a TCP connection, e.g., to introduce packet loss or to emulate a congestion situation.

The official NIST Net website is http://snad.ncsl.nist.gov/itg/nistnet/install.html. To install NIST Net, do the following.

1. Recompile the Linux kernel[4]. When configuring the kernel using **make menu-configure**, do the following:
   (a) Turn off module versioning: disable the "Loadable module support" → "Set version information on all module symbols" option.
   (b) Set the realtime clock driver to be compiled as a module: Select "M" for the "Character device"→"Enhanced Real Time Clock Support" option.
2. Execute **tar xvfz nistnet.2.0.12.tar.gz** to extract the package into a directory, e.g., /home/guest/nistnet/.
3. Execute **make**, and **make install**.
4. Append the directory of the NIST Net executables to the PATH environment variable of guest.

The Netspy multicast tool

Netspy is a simple multicast tool written by one of the authors of this book, Shiwen Mao, for the multicast exercises in Chapter 7. It has two components: **netspy**, which is a multicast client, and **netspyd**, which is a multicast sender. The source code for these two programs is given in Appendix C.2. To compile the code on Red Hat Linux, use:

gcc -o netspy netspy.c -lnsl -lresolv and
gcc -o netspyd netspyd.c -lnsl -lresolv.

---

[4] See the http://www.tldp.org/HOWTO/Kernel-HOWTO/ page on how to compile a Linux kernel.

To compile the code on Solaris 8.0, an additional compilation option **-lsocket** is needed.

The executables can be put in the `/usr/local/netspy` directory, and this directory should be appended to the `PATH` environment variable.

Accessory files

The TFTP exercise in Chapter 5 uses two files with randomly generated contents, a small file (1 kbyte) called `small.dum` and a large file (1 Mbyte) called `large.dum`. You can put these two files in the `/tftpboot` directory. The FTP exercise in Chapter 5 also uses these two files, but stored in a different directory at `/home/LAB/`.

The DBS exercises in Chapter 6 uses three command files in the `/home/guest/` directory: TCP1.cmd, TCP2.cmd, and TCPUDP.cmd. The files are given in Appendix C.1. Also a directory named `/home/guest/data` should be created to store data files for the DBS experiments.

You need a video clip for the realtime multicasting exercises in Chapter 7. You can download the video clip from, e.g., `http://www.gomovietrailers.com`, and put it in the `/home/guest/` video.

The Apache exercises in Chapter 8 uses two HTML files given in Appendix C.3. These two HTML files, along with an arbitrary GIF formatted picture file `mypic.gif` are stored in the `/var/www/html/` directory. A Perl CGI script, `hello.pl`, which is given in Appendix C.3, should be stored in the `/var/www/cgi-bin` directory. Note that you need to run **chmod +x hello.pl** to make it executable.

The four C programs for the socket programming exercises in Chapter 8 are given in Appendix C.4. These four files may be stored in the `/home/guest` directory.

A.3.3 Router configuration

In order to configure a router, you need to access the router from a computer, then run the configuration commands from the router. There are two ways to connect to a router. If you know the `login` and `enable` passwords, you can **telnet** to the router through one of its Ethernet interfaces. See Chapter 3 for a detailed introduction to the Cisco IOS and on how to configure a router through **telnet**.

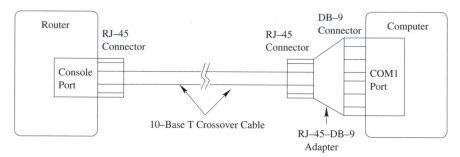

Figure A.2. Connecting to the router's console port.

Table A.2. *Default IP and MAC addresses for the routers*

router	Label	IP Address	MAC Address
1	router1 eth0	128.238.61.1	
	router1 eth1	128.238.62.1	
2	router2 eth0	128.238.62.2	
	router2 eth1	128.238.63.2	
3	router3 eth0	128.238.63.3	
	router3 eth1	128.238.64.3	
4	router4 eth0	128.238.64.4	
	router4 eth1	128.238.65.4	

Another way to access a router is through its `console` interface using a crossover cable. As shown in Fig. A.2, you can use a crossover 10 Base-T cable with RJ-45 connectors and a RJ-45-to-DB-9 adapter to connect the router console port to the COM1 port of the computer. Then you can run a terminal emulator (e.g., `c-kermit` from `http://www.columbia.edu/ kermit/ckermit.html`, or HyperTerminal in Microsoft Windows) to access the router. The serial connection should have a data rate of 9600 baud, 1 stop bit with 8 data bits, no flow control, and no parity.

Once you access the router, you can begin to configure it. See Chapter 3 for instruction on configuring a Cisco router. More details can be found from the Cisco website `http://www.cisco.com/`. If you do not know either the `login` or `enable` password, refer to the document `http:// www.cisco.com/warp/public/474/pswdrec_2600.shtml` on how to recover it.

The default IP addresses of the router interfaces are given in Table A.2. The router default configuration is given in Appendix B. As in Table A.1, you need to insert the MAC addresses of the router interfaces, which you can find by running **tcpdump** in the lab or from the router documentations, in Table A.2.

A.4 Estimated budget

All the diagnostic tools we use in the experiments are public domain software, which can be downloaded free of charge. For the hardware devices, although this guide is based on a setup of eight computers and four routers, a smaller setup is adequate. Table A.3 lists the minimum hardware requirements for each lab.

Also note that the experiments focus on the networking protocol operations, and thus do not require high-end computers and routers. Our experience shows that PCs with middle of range (or low) configuration (e.g., PII 500MHZ CPU, 256M memory, and several hundred MB of harddrive) or Sun Ultra 5 workstations, and Cisco 2600 series routers are adequate for all the exercises. The estimated budget for a 8-seat laboratory is about $17,500, which consists of $4 \times \$2000$ for the routers, $8 \times \$1000$ for the computers, and $1500 for accessories (hubs, cables, connectors, desks, and chairs). Further reduction on cost can be achieved if second-hand computers and routers are available or if the minimum setup shown in Table A.3 is used.

Table A.3. *The minimum set of workstations and routers needed for each lab*

Laboratory	Number of Workstations	Number of Routers
Chapter 1	2	0
Chapter 2	3	1
Chapter 3	2	2
Chapter 4	3	2
Chapter 5	2	1
Chapter 6	2	0
Chapter 7	3	2
Chapter 8	5	1
Chapter 9	2	0

A.5 Root privilege for system commands

We would suggest the creation of a normal user account on each host with the same user ID and password for students to use, e.g., user "guest" with password "guest1". Allowing students to login as "root" is not recommended, especially when some of them are not familiar with Linux or Unix systems. However, when configuring a host, students still need root privilege to run several system tools which are only executable by the root. There are several ways to grant partial root privilege to a normal user, as discussed in the following.

A.5.1 The wheel group

In most Linux systems, there is a preinstalled `wheel` group. By adding a user to the `wheel` group and making a certain system command accessible to the `wheel` group, the administrator can have flexible control over assigning the root privileges. The general procedure is the following.

1. Edit /etc/group, add the target user's id, e.g. `guest`, in the wheel group, as:

```
wheel:x:10:root,guest
```

2. Change the group attribute of the executable, e.g. `/sbin/xyz`, to the wheel group: **chgrp wheel /sbin/xyz**.
3. Change the mode of the executable, using **chmod 4750 /sbin/xyz**.

Note that this approach does not work for **ethereal**, since both `setuid` and `setgid` are not supported by GTK+ for security considerations. A normal user can execute the **ethereal** program, but he or she is not allowed to capture packets from the network interfaces. In the following experiments, we will use **tcpdump** to capture a packet trace first, then use **ethereal** to load the trace and analyze it.

A.5.2 Role-Based Access Control

Role-Based Access Control (RBAC) is a standard of National Institute of Standards and Technology (NIST) aiming to assign limited administrative capabilities to normal users. Sun's Solaris 8.0 is one of its first commercial implementations. RBAC is so popular that there has been an annual ACM RBAC workshop since 1996. The official RBAC website is: `http://csrc.nist.gov/rbac/`.

RBAC uses a set of configuration files to define a number of roles and profiles, and to assign a specific capability to a certain role. An interested reader can refer to [18] for more details.

A.5.3 Sudo

Sudo is another useful tool that allows an administrator to assign privilege for some users to run some commands as root. It also logs user behavior for auditing purposes. Sudo is preinstalled in Red Hat Linux 9.0.

To assign a command (e.g., **tcpdump**) to a normal user (e.g. guest), you need to edit the Sudo configuration file `/etc/sudoers` using the Sudo editor `visudo`, i.e., adding a new line at the end of the `/etc/sudoers` file:

```
guest    ALL=/usr/sbin/tcpdump
```

Note that it is a TAB character after guest in the above line. Then user guest can run **tcpdump** using:
sudo /usr/sbin/tcpdump.
He or she will be prompted for user guest's password. After entering the correct password, **tcpdump** begins to run. In the next five minutes, guest can run **tcpdump** without entering a password. After that, the user will be prompted for the password again.

Sudo is not preinstalled in Solaris 8.0. You can download the package from `http://www.courtesan.com/sudo/`. To install Sudo on a Solaris machine:

1. Download the Sudo source package, `sudo-1.6.3.tar.gz`.
2. Uncompress and extract the source package, by **gunzip sudo-1.6.3.tar.gz** and **tar xvf sudo-1.6.3.tar**.
3. Get into the `../sudo-1.6.3` directory. Execute: **./configure**, **make**, and **make install**.

Sudo and the Sudo editor **visudo** are installed at `/usr/local/bin`. You may need to append these directories into the `PATH` environment variable of user guest.

A.6 Internet access

All experiments in this guide can be done without Internet access. In fact, to avoid disturbing the normal operation of the campus network, it is recommended to keep the experimental network isolated. Students need to use floppy disks or other storage media to copy the experimental data.

If Internet access is available in the lab, students can transfer the experimental data using FTP or email. For example, there could be a ninth workstation which is multihomed, as shown in Fig. A.1. One interface of this machine connects to the internal network, while the other interface connects to the Internet. Note that routing and forwarding should be disabled in this multihomed host to keep the laboratory network isolated. We need to make sure that the Linux IP Forwarding module is not loaded, by running the following command:

echo "0" > **/proc/sys/net/ipv4/ip_forward**.

In Solaris, login as the root and execute: **touch /etc/notrouter**, to create an empty file /etc/notrouter. When the machine reboots, the startup script looks for the presence of this file. If it exists, the startup script will not start **in.routed -s** or **in.rdisc -r**, and does not turn on IP Forwarding.

Appendix B: initial configuration of the routers

B.1 Initial configuration of router1

Configuring global parameters.

```
Enter host name [router1]: router1
Enter enable password : e1537
Enter virtual terminal password : e1537
Configure SNMP Network Management? [yes]: no
Configure IP? [yes]: yes
  Configure IGRP routing? [no]: no
  Configure RIP routing? [yes]: no
Configure DECnet? [no]: no
Configure XNS? [no]: no
Configure Novell? [no]: no
Configure AppleTalk? [no]: no
Configure CLNS? [no]: no
Configure Vines? [no]: no
Configure bridging? [no]: no
```

Configuring interface parameters.

```
Configuring interface Ethernet0:
  Is this interface in use? [yes]: yes
  Configure IP on this interface? [yes]: yes
    IP address for this interface : 128.238.61.1
    Number of bits in subnet field [0]: 8
    Class B network is 128.238.0.0,
    8 subnet bits; mask is 255.255.255.0
```

```
Configuring interface Ethernet1:
  Is this interface in use? [yes]: yes
  Configure IP on this interface? [yes]: yes
    IP address for this interface : 128.238.62.1
    Number of bits in subnet field [0]: 8
    Class B network is 128.238.0.0,
    8 subnet bits; mask is 255.255.255.0
```

The following configuration command script was created.

```
hostname router
enable password el537
line vty 0 4
password el537
!
ip routing
no decnet routing
no xns routing
no novell routing
no appletalk routing
no clns routing
no vines routing
no bridge 1
!
interface Ethernet0
ip address 128.238.61.1 255.255.255.0
!
interface Ethernet1
ip address 128.238.62.1 255.255.255.0
!

Use this configuration? [yes/no]: yes
[OK]
Use the enabled mode 'configure' command
to modify this configuration.
```

B.2 Initial configurations of the other routers

The initial configurations of the other routers are similar to that of `router1`, with different interface IP addresses, as given in Table A.2.

Appendix C: source code

C.1 Command files for the DBS experiments

The following two command files are modified from the examples in the
DBS package.

C.1.1 TCP1.cmd

This is the command file for a single TCP connection.

```
# TCP1
{
        sender {
                hostname  = host2;
                port      = 0;
                so_debug  = OFF;
                tcp_trace = OFF;
                no_delay  = OFF;
                send_buff = 32768;
                recv_buff = 32768;
                mem_align = 2048;
                pattern {8192,    8192,    0.0,    0.0}
        }
        receiver {
                hostname  = host1;
                port      = 20000;
                so_debug  = OFF;
                tcp_trace = OFF;
                no_delay  = OFF;
                recv_buff = 32768;
                send_buff = 32768;
                mem_align = 8192;
                pattern {8192,    8192,    0.0,    0.0}
        }
        file        = data/tcp1-host2-host1;
        protocol    = TCP;
        start_time  = 0.0;
        end_time    = 30;
        send_times  = 2048;
}
```

C.1.2 TCPUDP.cmd

This is the command file for a single TCP connection and a UDP flow.

```
# TCP
{
        sender {
                hostname  = host2;
                port      = 0;
                so_debug  = OFF;
                tcp_trace = OFF;
                no_delay  = OFF;
                send_buff = 32768;
                recv_buff = 32768;
                mem_align = 2048;
                pattern {8192,    8192,     0.0,       0.0}
        }
        receiver {
                hostname  = host1;
                port      = 20000;
                so_debug  = OFF;
                tcp_trace = OFF;
                no_delay  = OFF;
                recv_buff = 32768;
                send_buff = 32768;
                mem_align = 8192;
                pattern {8192,    8192,     0.0,       0.0}
        }
        file            = data/tcpudp-host2-host1;
        protocol        = TCP;
        start_time      = 0.0;
        end_time        = 30;
        send_times      = 2048;
}

# UDP
{
        sender {
                hostname  = host3;
                port      = 20000;
                so_debug  = OFF;
                tcp_trace = OFF;
                no_delay  = OFF;
                send_buff = 32768;
                recv_buff = 32768;
                mem_align = 8192;
                # A sample MPEG traffic paterm (GOP=12)
                # This is modeled from
                # Craig Partridge, Gigabit Networking,
                #          Addison-Wesley, p187, 1993
                pattern {40960, 1024,  0.033333333,  0;
                          2048, 1024,  0.033333333,  0;
                          2048, 1024,  0.033333333,  0;
                         10240, 1024,  0.033333333,  0;
                          2048, 1024,  0.033333333,  0;
                          2048, 1024,  0.033333333,  0;
                         10240, 1024,  0.033333333,  0;
```

```
                                    2048,  1024,  0.033333333,  0;
                                    2048,  1024,  0.033333333,  0;
                                   10240,  1024,  0.033333333,  0;
                                    2048,  1024,  0.033333333,  0;
                                    2048,  1024,  0.033333333,  0;}
                }
        receiver {
                hostname  = host1;
                port      = 20000;
                mem_align = 2048;
                pattern {8192,  8192,    0.0,     0.0}
        }
        file            = data/tcpudp-host3-host1;
        protocol        = UDP;
        start_time      = 2.0;
        end_time        = 30;
        send_times      = 50;
}
```

C.1.3 TCP2.cmd

This is the command file for two TCP connections.

```
# TCP1
{
        sender {
                hostname  = host2;
                port      = 0;
                so_debug  = OFF;
                tcp_trace = OFF;
                no_delay  = OFF;
                send_buff = 32768;
                recv_buff = 32768;
                mem_align = 2048;
                pattern {8192,  8192,    0.0,     0.0}
        }
        receiver {
                hostname  = host1;
                port      = 20000;
                so_debug  = OFF;
                tcp_trace = OFF;
                no_delay  = OFF;
                recv_buff = 32768;
                send_buff = 32768;
                mem_align = 8192;
                pattern {8192,  8192,    0.0,     0.0}
        }
        file            = data/tcp2-host2-host1;
        protocol        = TCP;
        start_time      = 0.0;
        end_time        = 30;
        send_times      = 2048;
}
# TCP2
{
        sender {
```

```
                         hostname  = host3;
                         port      = 0;
                         so_debug  = OFF;
                         tcp_trace = OFF;
                         no_delay  = OFF;
                         send_buff = 32768;
                         recv_buff = 32768;
                         mem_align = 2048;
                         pattern {8192,  8192,    0.0,     0.0}
                }
        receiver {
                         hostname  = host1;
                         port      = 20000;
                         so_debug  = OFF;
                         tcp_trace = OFF;
                         no_delay  = OFF;
                         recv_buff = 32768;
                         send_buff = 32768;
                         mem_align = 8192;
                         pattern {8192,  8192,    0.0,     0.0}
                }
        file            = data/tcp2-host3-host1;
        protocol        = TCP;
        start_time      = 0.0;
        end_time        = 30;
        send_times      = 2048;
}
```

C.2 Netspy source code

C.2.1 netspy.c source code

```
/******************************************/
/* netspy.c                               */
/*                                        */
/* Written by Dr Shiwen Mao               */
/* Dept. ECE, Polytechnic University      */
/*                                        */
/* 7/17/2003                              */
/******************************************/

#include <sys/types.h>
#include <sys/socket.h>
#include <netinet/in.h>
#include <arpa/inet.h>
#include <netdb.h>
#include <stdio.h>
#include <unistd.h> /* close */

#define MAX_MSG 100

void main(int argc, char *argv[]) {

    int sd, rc, n, cliLen;
    struct ip_mreq mreq;
```

```c
struct sockaddr_in cliAddr, servAddr;
struct in_addr mcastAddr;
struct hostent *h;
char msg[MAX_MSG];

int  SERVER_PORT;

if(argc!=3) {
  printf("usage : %s <mcast address> <mcast
    port number>\n", \
                    argv[0]);
  exit(0);
}

SERVER_PORT = atoi(argv[2]);

/* get mcast address to listen to */
h=gethostbyname(argv[1]);
if(h==NULL) {
  printf("%s : unknown group '%s'\n",argv[0],argv[1]);
  exit(1);
}

memcpy(&mcastAddr, h->h_addr_list[0],h->h_length);

/* check given address is multicast */
if(!IN_MULTICAST(ntohl(mcastAddr.s_addr))) {
  printf("%s : given address '%s' is not multicast\n",
    argv[0], \
 inet_ntoa(mcastAddr));
  exit(1);
}

/* create socket */
sd = socket(AF_INET,SOCK_DGRAM,0);
if(sd<0) {
  printf("%s : cannot create socket\n",argv[0]);
  exit(1);
}

/* bind port */
servAddr.sin_family=AF_INET;
servAddr.sin_addr.s_addr=htonl(INADDR_ANY);
servAddr.sin_port=htons(SERVER_PORT);
if(bind(sd,(struct sockaddr *) &servAddr,
    sizeof(servAddr))<0){
  printf("%s : cannot bind port %d \n",argv[0],
    SERVER_PORT);
  exit(1);
}

/* join multicast group */
mreq.imr_multiaddr.s_addr=mcastAddr.s_addr;
mreq.imr_interface.s_addr=htonl(INADDR_ANY);

rc = setsockopt(sd,IPPROTO_IP,IP_ADD_MEMBERSHIP, \
```

```
                                         (void *) &mreq,
                                            sizeof(mreq));
      if(rc<0) {
        printf("Netspy : cannot join multicast group
         '%s'", \
                            inet_ntoa(mcastAddr));
        exit(1);
      }
      else {
        printf("\nNetspy : listening to mgroup %s:%d\n\n", \
                            inet_ntoa(mcastAddr),
                                SERVER_PORT);

        /* infinite server loop */
        while(1) {
          cliLen=sizeof(cliAddr);
          n = recvfrom(sd,msg,MAX_MSG,0, \
                            (struct sockaddr *) &cliAddr,
                                &cliLen);
          if(n<0) {
            printf("Netspy : cannot receive data\n");
            continue;
          }

          printf("    == : %s\n", msg);
        }/* end of infinite server loop */
      }
    }
```

C.2.2 netspyd.c source code

```
/****************************************/
/* netspyd.c                            */
/*                                      */
/* Written by Dr Shiwen Mao             */
/* Dept. ECE, Polytechnic University    */
/*                                      */
/* 7/17/2003                            */
/****************************************/

#include <sys/types.h>
#include <sys/socket.h>
#include <netinet/in.h>
#include <arpa/inet.h>
#include <netdb.h>
#include <stdio.h>
#include <unistd.h> /* close */

#include <utmp.h>
#include <time.h>
#include <sys/timeb.h>
#include <string.h>
#include <netdb.h>

#define MAX_LOGIN 256

int main(int argc, char *argv[]) {
```

```
int sd, i;
unsigned char ttl;
struct sockaddr_in cliAddr, servAddr;
struct hostent *h;
int SERVER_PORT;
char ampm[]="AM";
struct tm* today;
char hostn[128];
char message[200];
struct utmp *upt_var;
struct utmp old_utmp[MAX_LOGIN];
int    flags[MAX_LOGIN];
int    old_len, found = -1;

old_len = 0;
for(i=0;i<MAX_LOGIN;i++){
  flags[i] = -1;
}

gethostname(hostn, sizeof(hostn));

if(argc!=4) {
  printf("usage netspyd <mgroup_addr> <port_number> \
                    <TTL_value>\n");
  exit(1);
}
ttl = atoi(argv[3]);

SERVER_PORT = atoi(argv[2]);

h = gethostbyname(argv[1]);
if(h==NULL) {
  printf("netspyd : unknown host '%s'\n", argv[1]);
  exit(1);
}

servAddr.sin_family = h->h_addrtype;
memcpy((char *) &servAddr.sin_addr.s_addr, \
                    h->h_addr_list[0],
                    h->h_length);
servAddr.sin_port = htons(SERVER_PORT);

if(!IN_MULTICAST(ntohl(servAddr.sin_addr.s_addr))) {
  printf("netspyd : address '%s' is not multicast
    \n", \
                    inet_ntoa(servAddr.sin_addr));
  exit(1);
}

sd = socket(AF_INET,SOCK_DGRAM,0);
if (sd<0) {
  printf("netspyd : cannot open socket\n");
  exit(1);
}

cliAddr.sin_family = AF_INET;
```

```
cliAddr.sin_addr.s_addr = htonl(INADDR_ANY);
cliAddr.sin_port = htons(SERVER_PORT+1);
if(bind(sd,(struct sockaddr *) &cliAddr,
   sizeof(cliAddr))<0) {
  perror("bind");
  exit(1);
}
if(setsockopt(sd,IPPROTO_IP,IP_MULTICAST_TTL,&ttl,
   sizeof(ttl))<0){
  printf("netspyd : cannot set ttl = %d \n",ttl);
  exit(1);
}

printf("netspyd started : \n");
printf("                   [local address   : %s:%d]
  \n" , \
   hostn, SERVER_PORT+1);
printf("                   [multicast group : %s:%d]
  \n\n" , \
   inet_ntoa(*(struct in_addr *)h->h_addr_list[0]),
    SERVER_PORT);

while(1){
  setutent();

  while((upt_var = getutent()) != NULL){
    if(upt_var->ut_type == USER_PROCESS){
      found = -1;
      for(i=0;i<old_len;i++){
        if(upt_var->ut_pid == old_utmp[i].ut_pid){
          found = 1;
          flags[i] = 1;
        }
      }

      if(found == -1){
        today = localtime(&upt_var->ut_time);
        if(today->tm_hour>12){
          ampm[0] = 'P';
          today->tm_hour -= 12;
        }
        else{
          ampm[0] = 'A';
        }

        printf("  == : %s logged on to %s at %.5s %s,
          pid=%d\n",\
                upt_var->ut_user,hostn, asctime(today)
                  +11, ampm, \
                upt_var->ut_pid);
        sprintf(message, "%s logged on to %s at %.5s
          %s", \
                upt_var->ut_user,hostn, asctime(today)
                  +11, ampm);
```

```
                  sendto(sd, message, strlen(message)+1, 0, \
                        (struct sockaddr *) &servAddr, sizeof
                        (servAddr));
                }
            }
        }

        endutent();

        for(i=0;i<old_len;i++){
          if(flags[i] == -1){
            today = localtime(&old_utmp[i].ut_time);
            if(today->tm_hour>12){
              ampm[0] = 'P';
              today->tm_hour -= 12;
            }
            else{
              ampm[0] = 'A';
            }
            printf("  ==  : %s logged out from %s at %.5s %s,
               pid=%d\n",\
                     old_utmp[i].ut_user, hostn, asctime(today)
                       +11, ampm,\
                     old_utmp[i].ut_pid);
            sprintf(message, "%s logged out from %s at %.5s
               %s", \
                     old_utmp[i].ut_user,hostn, asctime(today)
                       +11, ampm);
            sendto(sd, message, strlen(message)+1, 0, \
                        (struct sockaddr *) &servAddr, sizeof
                        (servAddr));
          }
        }

        setutent();

        old_len = 0;
        while((upt_var = getutent()) != NULL){
          if(upt_var->ut_type == USER_PROCESS){
            memcpy((struct utmp *)&old_utmp[old_len], \
                                 upt_var,
                                   sizeof(struct utmp));
            flags[old_len] = -1;
              old_len = old_len + 1;
          }
        }
        endutent();
        sleep(2);
    }

    close(sd);
    exit(0);
}
```

C.3 HTML and CGI files

C.3.1 The `try1.html` file used in Chapter 8

This is an example HTML file, with a line of text, a hyperlink, and an embedded picture.

```
<html>
<head>
<title>An Example HTML File</title>
</head>
<body>
<p>This is a text line.</p>
<p><a href="usage/index.html">This is a hyperlink</a>.
   </p>
<p>Here is an embedded picture:
<img border="0" src="mypic.gif" width="164"
   height="123"></p>
</body>
</html>
```

C.3.2 The `try2.html` file used in Chapter 8

This is a HTML form, with which a user can send data to the web server, and to invoke the CGI script on the server to handle the data.

```
<html>
<head>
<TITLE>An Exsample HTML FORM</TITLE>
</head>

<body>
<p>
<p>
<hr>
<FORM ACTION="/cgi-bin/hello.pl" METHOD="GET">
Type you name:
<p>    <INPUT TYPE="TEXT" NAME="name">
<p>    <INPUT TYPE="SUBMIT">
</FORM>
<hr>
</body>
</html>
```

C.3.3 The `hello.pl` CGI script used in Chapter 8

This CGI script reads a text string, and returns a HTML file with the line "This is the data you entered: *string*!". The script is written in Perl.

```perl
#!/usr/bin/perl

print "Content-type: text/html\r\n\r\n";
print "<html><head><title>CGI Response</title>
   </head>\r\n";
print "<hr>\r\n";
print "<p>\r\n";
print "The data received by the server is:\r\n";
print "<p>\r\n";
print "$ENV{'QUERY_STRING'}\r\n";
print "<p>\r\n";
print "<hr>\r\n";
print "</body></html>\r\n";
```

In Solaris, change the first line to "#!/usr/local/bin/perl".

C.4 Socket programming source codes

C.4.1 UDPserver.c

```c
//////////////////////////////////////////////////////////
// UDPserver.c -- Sockets that use UDP datagrams      //
//                                                    //
// Written by Dr Shiwen Mao, Polytechnic Univ.        //
// December 2003.                                     //
//////////////////////////////////////////////////////////
  #include <stdio.h>
  #include <stdlib.h>
  #include <unistd.h>
  #include <string.h>
  #include <sys/types.h>
  #include <sys/socket.h>
  #include <netinet/in.h>
  #include <arpa/inet.h>

  #define BUFFLEN 100

  int main(int argc, char* argv[])
  {
    int      sockserver;
    struct sockaddr_in server_addr;
    struct sockaddr_in client_addr;
    int      addr_len, sendlen, rcvdlen;
    char     buf[BUFFLEN];

    if(argc !=2){
       printf("Usage: UDPserver server_port \n");
       exit(1);
    }
```

```c
// Create the UDP server socket
if ((sockserver = socket(AF_INET, SOCK_DGRAM,
        0))==-1){
    printf("Error in creating UDP socket.\n");
    exit(1);
}

// Set the server socket address
server_addr.sin_family = AF_INET;
server_addr.sin_port = atoi(argv[1]);
server_addr.sin_addr.s_addr = INADDR_ANY;
memset(&(server_addr.sin_zero), '\0', 8);

// Associate the server socket address with the
    server socket
if (bind(sockserver, (struct sockaddr *)
    &server_addr, \
                            sizeof(struct sockaddr))
                                ==-1){
    printf("Error in binding the socket address.\n");
    exit(1);
}

while(1){
    // Receive a message from the UDP client socket
    addr_len = sizeof(struct sockaddr);
    if ((rcvdlen=recvfrom(sockserver,buf, BUFFLEN-1,
        0, \
      (struct sockaddr *)&client_addr, &addr_len))
        == -1) {
        printf("Error in recvfrom.\n");
        exit(1);
    }
    printf("got a %i byte packet from client %s\n", \
                    rcvdlen,inet_ntoa
                    (client_addr.sin_addr));
    buf[rcvdlen] = '\0';
    printf("message: %s\n",buf);

    // Return the message to the client
    if ((sendlen=sendto(sockserver, buf, rcvdlen, 0, \
                    (struct sockaddr *)&client_addr, \
                        sizeof(struct sockaddr)))==-1){
        printf("Error in sendto.\n");
        exit(1);
    }
    printf("returned %d bytes to client %s\n",
        sendlen,\
                            inet_ntoa
                            (client_addr.sin_addr));
    printf("message: %s\n\n", buf);
}

return 0;
}
```

C.4.2 UDPclient.c

```
/////////////////////////////////////////////////////
// UDPclient.c -- Sockets that use UDP datagrams    //
//                                                  //
// Written by Dr Shiwen Mao, Polytechnic Univ.      //
// December 2003.                                   //
/////////////////////////////////////////////////////

#include <stdio.h>
#include <stdlib.h>
#include <unistd.h>
#include <string.h>
#include <sys/types.h>
#include <sys/socket.h>
#include <netinet/in.h>
#include <arpa/inet.h>
#include <netdb.h>

#define BUFFLEN 100

int main(int argc, char *argv[])
{
    int     sockclient;
    struct sockaddr_in server_addr;
    struct hostent *hent;
    int     sendlen,rcvdlen,addrlen;
    char    buf[BUFFLEN];

    if (argc != 4) {
        printf("Usage: UDPclient server_ip server_port
         message\n");
        exit(1);
    }

    // Get the UDP server's IP address
    if ((hent=gethostbyname(argv[1])) == NULL) {
        printf("Error in gethostbyname.\n");
        exit(1);
    }

    // Create the UDP client socket
    if ((sockclient = socket(AF_INET, SOCK_DGRAM, 0))
        == -1) {
        printf("Error in creating UDP socket.\n");
        exit(1);
    }

    // Set the UDP server's address
    server_addr.sin_family = AF_INET;
    server_addr.sin_port = atoi(argv[2]);
    server_addr.sin_addr = *((struct in_addr *)
     hent->h_addr);
    memset(&(server_addr.sin_zero), '\0', 8);

    // Send the message to the UDP server
    if ((sendlen=sendto(sockclient, argv[3],
```

```
            strlen(argv[3]), \
                            0, (struct sockaddr *)
                            &server_addr, \
                            sizeof(struct sockaddr)))
                            ==-1){
        printf("Error in sendto.\n");
        exit(1);
    }
    printf("sent %d bytes to %s\n", \
                sendlen, inet_ntoa
                (server_addr.sin_addr));
    printf("message: %s\n\n", argv[3]);

    // Receive the returned message from the server
    if ((rcvdlen=recvfrom(sockclient,buf, BUFFLEN-1,
      0, \
        (struct sockaddr *)&server_addr, &addrlen))
        == -1){
        printf("Error in recvfrom.\n");
        exit(1);
    }
    printf("received %d bytes from server %s\n",
      rcvdlen, \
                        inet_ntoa
                        (server_addr.sin_addr));
    buf[rcvdlen] = '\0';
    printf("received message: %s\n", buf);

    // Shutdown the UDP client socket
    close(sockclient);

    return 0;
}
```

C.4.3 TCPserver.c

```
/////////////////////////////////////////////////////
// TCPserver.c -- A TCP server socket              //
//                                                 //
// Written by Dr Shiwen Mao, Polytechnic Univ.     //
// December 2003.                                  //
/////////////////////////////////////////////////////
#include <stdio.h>
#include <stdlib.h>
#include <unistd.h>
#include <string.h>
#include <sys/types.h>
#include <sys/socket.h>
#include <netinet/in.h>
#include <arpa/inet.h>

#define BACKLOG 10
#define BUFFLEN 100

int main(int argc, char *argv[])
```

```c
{
    int     sockserver, sockclient;
    struct sockaddr_in server_addr, client_addr;
    int     sockin_size;
    int     sendlen, rcvdlen;
    char    buf[BUFFLEN];

    if (argc != 2){
        printf("Usage: TCPserver server_port\n");
        exit(1);
    }

    // Create the TCP server socket
    if ((sockserver = socket(AF_INET, SOCK_STREAM, 0))
     == -1) {
        printf("Error in creating the server socket.\n");
        exit(1);
    }

    // Set the server socket address
    server_addr.sin_family = AF_INET;
    server_addr.sin_port = atoi(argv[1]);
    server_addr.sin_addr.s_addr = INADDR_ANY;
    memset(&(server_addr.sin_zero), '\0', 8);

    // Associate the server address with the
        server socket
    if (bind(sockserver, (struct sockaddr *)
     &server_addr, \
            sizeof(struct sockaddr)) == -1) {
        printf("Error in bind.\n");
        exit(1);
    }

    // Waiting for client requests
    if (listen(sockserver, BACKLOG) == -1) {
        printf("Error in listen.\n");
        exit(1);
    }

    while(1) {
        // Accept a client connection request
        sockin_size = sizeof(struct sockaddr_in);
        if ((sockclient = accept(sockserver, \
            (struct sockaddr *)&client_addr, \
                    &sockin_size)) == -1) {
            printf("Error in accept.\n");
            continue;
        }
        printf("TCP server: connection request from
        %s\n", \
                        inet_ntoa
                        (client_addr.sin_addr));

        // Receive a message from the connected client
        if ((rcvdlen=recv(sockclient, buf, BUFFLEN-1, 0))
         ==-1){
```

```
                    printf("Error in recv.\n");
                    continue;
                  }
                  buf[rcvdlen] = '\0';
                  printf("Received from client: %s\n", buf);

                  // Return the message to the client
                  if ((sendlen=send(sockclient, buf, rcvdlen, 0))
                    == -1){
                    printf("Error in send.\n");
                    continue;
                  }
                  printf("Sent to client: %s\n\n", buf);

                  // Close the client socket, terminate the TCP
                      connection.
                  close(sockclient);
                }

                return 0;
              }
```

C.4.4 TCPclient.c

```
///////////////////////////////////////////////////////
// TCPclient.c -- A TCP client socket                 //
//                                                    //
// Written by Dr Shiwen Mao, Polytechnic Univ.        //
// December 2003.                                     //
///////////////////////////////////////////////////////
#include <stdio.h>
#include <stdlib.h>
#include <unistd.h>
#include <string.h>
#include <netdb.h>
#include <sys/types.h>
#include <netinet/in.h>
#include <sys/socket.h>

#define BUFFLEN 100

int main(int argc, char *argv[])
{
  int sockserver;
  struct hostent *hent;
  struct sockaddr_in server_addr;
  int     sendlen, rcvdlen;
  char    buf[BUFFLEN];

  if (argc != 4) {
    printf("Usage: TCPclient server_ip server_port
     message\n");
    exit(1);
  }

  // Get the TCP server's IP address
```

1042   Standard for the Transmission of IP Datagrams over IEEE 802 Networks, J. Postel, J. K. Reynolds [Feb-01-1988].

1058   Routing Information Protocol, C. L. Hedrick [Jun-01-1988].

1071   Computing the Internet Checksum, R. T. Braden, D. A. Borman, C. Partridge [Sep-01-1988].

1108   US Department of Defense Security Options for the Internet Protocol, S. Kent [November 1991].

1112   Host Extensions for IP Multicasting, S. E. Deering [Aug-01-1989].

1122   Requirements for Internet Hosts – Communication Layers, R. Braden (ed.) [October 1989].

1123   Requirements for Internet Hosts – Application and Support, R. Braden (ed.) [October 1989].

1141   Incremental Updating of the Internet Checksum, T. Mallory, A. Kullberg [Jan-01-1990].

1157   Simple Network Management Protocol (SNMP), J. D. Case, M. Fedor, M. L. Schoffstall, J. Davin [May-01-1990].

1191   Path MTU Discovery, J. C. Mogul, S. E. Deering [Nov-01-1990].

1256   ICMP Router Discovery Messages, S. Deering (ed.) [Sep-01-1991].

1267   Border Gateway Protocol 3 (BGP-3), K. Lougheed, Y. Rekhter [Oct-01-1991].

1305   Network Time Protocol (Version 3) Specification, Implementation, D. Mills [March 1992].

1323   TCP Extensions for High Performance, V. Jacobson, R. Braden, D. Borman [May 1992].

1340   Assigned Numbers, J. Reynolds, J. Postel [July 1992].

1372   Telnet Remote Flow Control Option, C. Hedrick, D. Borman [October 1992].

1379   Extending TCP for Transactions – Concepts, R. Braden [November 1992].

1393   Traceroute Using an IP Option, G. Malkin [January 1993].

1441   Introduction to Version 2 of the Internet-standard Network Management Framework, J. Case, K. McCloghrie, M. Rose, S. Waldbusser [April 1993].

1454   Comparison of Proposals for Next Version of IP, T. Dixon [May 1993].

1480   The US Domain, A. Cooper, J. Postel [June 1993].

1519   Classless Inter-Domain Routing (CIDR): an Address Assignment and Aggregation Strategy, V. Fuller, T. Li, J. Yu, K. Varadhan [September 1993].

1534   Interoperation Between DHCP and BOOTP, R. Droms [October 1993].

1585   MOSPF: Analysis and Experience, J. Moy [March 1994].

1614   Network Access to Multimedia Information, C. Adie [May 1994].

1644   T/TCP – TCP Extensions for Transactions Functional Specification, R. Braden [July 1994].

1651   SMTP Service Extensions, J. Klensin, N. Freed, M. Rose, E. Stefferud, D. Crocker [July 1994].

1812   Requirements for IP Version 4 Routers, F. Baker (ed.) [June 1995].

1851   The ESP Triple DES Transform, P. Karn, P. Metzger, W. Simpson [September 1995].

2050   Internet Registry IP Allocation Guidelines, K. Hubbard, M. Kosters, D. Conrad, D. Karrenberg, J. Postel [November 1996].

2131   Dynamic Host Configuration Protocol, R. Droms [March 1997].

2178   OSPF Version 2, J. Moy [July 1997].

2181   Clarifications to the DNS Specification, R. Elz, R. Bush [July 1997].

2189   Core Based Trees (CBT version 2) Multicast Routing – Protocol Specification, A. Ballardie [September 1997].

2201   Core Based Trees (CBT) Multicast Routing Architecture, A. Ballardie [September 1997].

2236   Internet Group Management Protocol, Version 2, W. Fenner [November 1997].

2326   Real Time Streaming Protocol (RTSP), H. Schulzrinne, A. Rao, R. Lanphier [April 1998].

2453   RIP Version 2, G. Malkin [November 1998].

2474   Definition of the Differentiated Services Field (DS Field) in the IPv4 and IPv6 Headers, K. Nichols, S. Blake, F. Baker, D. Black [December 1998].

2594   Definitions of Managed Objects for WWW Services, H. Hazewinkel, C. Kalbfleisch, J. Schoenwaelder [May 1999].

2644   Changing the Default for Directed Broadcasts in Routers, D. Senie [August 1999].

2660   The Secure HyperText Transfer Protocol, E. Rescorla, A. Schiffman [August 1999].

2821	Simple Mail Transfer Protocol, J. Klensin (ed.) [April 2001].
2822	Internet Message Format, P. Resnick (ed.) [April 2001].
2908	The Internet Multicast Address Allocation Architecture, D. Thaler, M. Handley, D. Estrin [September 2000].
3022	Traditional IP Network Address Translator (Traditional NAT), P. Srisuresh, K. Egevang [January 2001].
3168	The Addition of Explicit Congestion Notification (ECN) to IP, K. Ramakrishnan, S. Floyd, D. Black [September 2001].
3170	IP Multicast Applications: Challenges and Solutions, B. Quinn, K. Almeroth [September 2001].
3228	IANA Considerations for IPv4 Internet Group Management Protocol (IGMP), B. Fenner [February 2002].
3306	Unicast-Prefix-based IPv6 Multicast Addresses, B. Haberman, D. Thaler [August 2002].
3307	Allocation Guidelines for IPv6 Multicast Addresses, B. Haberman [August 2002].
3397	Dynamic Host Configuration Protocol (DHCP) Domain Search Option, B. Aboba, S. Cheshire [November 2002].
3411	An Architecture for Describing Simple Network Management Protocol (SNMP) Management Frameworks, D. Harrington, R. Presuhn, B. Wijnen [December 2002].
3413	Simple Network Management Protocol (SNMP) Applications, D. Levi, P. Meyer, B. Stewart [December 2002].
3417	Transport Mappings for the Simple Network Management Protocol (SNMP), R. Presuhn (ed.) [December 2002].
3418	Management Information Base (MIB) for the Simple Network Management Protocol (SNMP), R. Presuhn (ed.) [December 2002].
3442	The Classless Static Route Option for Dynamic Host Configuration Protocol (DHCP) Version 4, T. Lemon, S. Cheshire, B. Volz [December 2002].
3447	Public-Key Cryptography Standards (PKCS) #1: RSA Cryptography Specifications Version 2.1, J. Jonsson, B. Kaliski [February 2003].
3489	STUN – Simple Traversal of User Datagram Protocol (UDP) Through Network Address Translators (NATs), J. Rosenberg, J. Weinberger, C. Huitema, R. Mahy [March 2003].

3493   Basic Socket Interface Extensions for IPv6, R. Gilligan, S. Thomson, J. Bound, J. McCann, W. Stevens [February 2003].

3512   Configuring Networks and Devices with Simple Network Management Protocol (SNMP), M. MacFaden, D. Partain, J. Saperia, W. Tackabury [April 2003].

3542   Advanced Sockets Application Program Interface (API) for IPv6, W. Stevens, M. Thomas, E. Nordmark, T. Jinmei [May 2003].

3559   Multicast Address Allocation MIB, D. Thaler [June 2003].

3600   Internet Official Protocol Standards, J. Reynolds (ed.), S. Ginoza (ed.) [November 2003].

3633   IPv6 Prefix Options for Dynamic Host Configuration Protocol (DHCP) Version 6, O. Troan, R. Droms [December 2003].

3646   DNS Configuration options for Dynamic Host Configuration Protocol for IPv6 (DHCPv6), R. Droms (ed.) [December 2003].

Index